Building Surveys

Building Surveys

Sixth edition

Peter Glover

AMSTERDAM • BOSTON • HEIDELBERG • LONDON • NEW YORK • OXFORD
PARIS • SAN DIEGO • SAN FRANCISCO • SINGAPORE • SYDNEY • TOKYO

Butterworth-Heinemann is an imprint of Elsevier

692

Butterworth-Heinemann is an imprint of Elsevier
Linacre House, Jordan Hill, Oxford OX2 8DP
30 Corporate Drive, Suite 400, Burlington, MA 01803

First published 1983, as *Surveying Buildings*
Second edition 1990, as *Building Surveys*
Third edition 1996
Reprinted 1998
Fourth edition 2001
Fifth edition 2003
Reprinted 2004
Sixth edition 2006

British Library Cataloguing in Publication Data
A catalogue record for this book is available from the British Library

Library of Congress Control Number: 2006925399

ISBN–13: 978-0-7506-8128-5
ISBN–10: 0-7506-8128-4

For information on all Butterworth-Heinemann publications
visit our website at http://books.elsevier.com

Typeset by Charon Tec Ltd, Chennai, India
www.charontec.com
Printed and bound in Italy
06 07 08 09 10   10 9 8 7 6 5 4 3 2 1

Working together to grow libraries in developing countries

www.elsevier.com | www.bookaid.org | www.sabre.org

ELSEVIER   BOOK AID International   Sabre Foundation

Contents

To Kay and the boys.
I am sure that they often wondered what
it was that I actually did for a living.

List of illustrations

process hastened by the electric motor and rotating disks which increase the exposure to oxygen in the air. The treated effluent emerging from the system is clear water suitable for draining into soakaways or an adjoining water course. Consent to discharge effluent is required from the Environment Agency. Periodic de-sludging is necessary.

Acknowledgements

My thanks to all the colleagues with whom I have worked over the years and fellow surveyors generally who have contributed to our knowledge of building performance and building defects, passing on their wisdom and experience in the process and to Nationwide Building Society which gave permission to reproduce a copy of its Mortgage Valuation Report form.

Preface

I left school at the age of 16 years and spent 5 years studying and following behind various surveyors as they prepared reports on buildings, finally qualifying as a Chartered Surveyor myself at the age of 21 years. Some of these surveyors were rather eccentric. One was accompanied on his inspections by a large shaggy sheepdog which he was trying to train to sniff out dry rot; I think that he was probably wasting his time. Another wrote his notes in undecipherable scribble on the back of the agents' sale particulars and then went back to the office to dictate a long rambling report, mostly from memory, including much information about plants in the garden and door knobs and hinges; clients in receipt of his reports were probably none the wiser as to whether there was actually anything wrong with the house.

As is usually the case with 21-year olds I then thought that I knew most of what I needed to know about Building Surveys and expected that the occasional curiosities or problems I would encounter would be easily resolved by means of quick recourse to one of the office text books or a chat with a colleague.

Buildings, and especially old buildings, can present a surprising variety of defects, however, and have many tricks waiting up their sleeves for the over-confident young surveyor. Clients, too, come with a wide range of requirements and expectations. And we are now in an age of consumerism when the customer requires comprehensive advice at minimum cost without delay and is more than happy to invoke the surveyor's complaints procedure or sue in the courts if actual or perceived shortcomings in the report are discovered.

Every year since I qualified I have discovered new or different aspects of the job and indeed as the years have passed there seems to be more, rather then less, to discover. So Building Surveys has been updated again to reflect the practical aspects of the work and proposed changes in the house selling process in England and Wales. Recent legislation and court decisions are also included.

I hope that the contents will be valued reading for all surveyors and related professionals and that members of the general public engaged in buying, owning or selling property will also find much useful information.

Introduction

I tell this tale, which is strictly true,
Just by way of convincing you
How very little, since things were made,
Things have altered in the building trade

    A Truthful Song, Rudyard Kipling

The preparation of a Building Survey Report provides an opportunity for the surveyor to fulfil two useful functions: First the client can be advised in relation to the condition of the building, the further investigations which may be needed and the advisability of the purchase; secondly by encouraging timely and sensible repairs the nation's building stock generally can be improved. So this is a useful service and a supply of well-trained surveyors using appropriate methodology is likely to be needed for the foreseeable future.

This book is written as a practical guide for surveyors and others who may be asked to survey buildings and prepare reports. It may also be of interest to members of the general public who may be contemplating buying or renovating buildings or who may be interested in the workings of the property market generally.

Since no two buildings, no two surveyors and no two clients will be the same the style and content of reports can vary considerably from case to case and in different localities. Certain broad principles of approach are advocated in this book, however.

First I would argue that reports should not be unnecessarily long – if there is nothing much to say on a particular topic keep it brief; surveyors are not helping their clients if they produce long reports with lots of padding, perhaps in an attempt to justify the fee. Secondly regard should be had to the Guidance Notes published by the Royal Institution of Chartered Surveyors where, amongst other things, great stress is placed on the need for surveyors to confirm the nature of their instructions in writing using appropriate terms and conditions of engagement to avoid misunderstandings and disputes with clients later. Thirdly the surveyor should adopt a methodology using site notes, field sheets and check lists so that there is a proper record of what was done and what was seen (and not seen) during the course of the inspection.

We live at a time when all the professions are subject to criticism and the general public are encouraged to complain and sue, moreover expectations in relation to the condition of the buildings we occupy and the workmanship of contractors are rising. Indeed if you watch daytime television (something the author avoids if he can) it would appear that our whole economy now consists of selling one another houses, doing them up and making personal accident claims.

All professionals are affected by this but there are few areas of activity where the potential for significant error giving rise to serious financial loss to the client can be as great as with a Building Survey. As we shall see in Chapter 19 hapless surveyors have often found themselves in the unhappy position of having to justify one of their reports in court and the duty of care to the client – and sometimes third parties as well – is considerable.

Alongside a tendency for surveyor's reports to become longer and more defensive we also have the increasing use of more abbreviated report forms such as the Homebuyer Report sponsored by the Royal Institution of Chartered Surveyors. Also the mortgage lending institutions routinely send a copy of their valuation reports to their customers (accompanied generally with warnings regarding the very limited nature of the mortgage valuation inspection). Some advice in relation to the completion of these reports is given in Chapter 11.

If all this were not enough the Government now proposes major changes to the house-buying process in England and Wales. Prior to placing a house (or flat) on the market the vendor will be required to prepare a Home Information Pack complete with a Home Condition Report prepared by a Home Inspector. Some advice in relation to this type of report is also included.

In case the reader may be young man or woman contemplating surveying as a career, or an older person considering a career change I would mention some of the attributes which seem to me to be needed for an individual undertaking this type of work. There are, of course, other branches of the surveying profession to which these comments may not necessarily apply.

First a surveyor should have a good command of the English language with an ability to produce well written reports and the facility, when necessary, to talk to clients in a helpful and constructive manner.

Secondly the surveyor should have a practical turn of mind. He or she should be able to understand the processes of building construction and the way in which repairs may be undertaken, how the various services work and be able to assimilate new ideas. It is helpful, although not essential, to have some personal experience of undertaking simple building operations such as basic carpentry, brickwork, plastering, wiring and plumbing and any time spent on site watching builders at work is always time well spent.

Thirdly the surveyor should have a reasonably tactful and pleasant personality. When inspecting houses in particular it will be necessary to visit properties, with the owners present, spending a considerable amount of time there with the intention of finding faults. The visit by the surveyor may be anxiously anticipated and the vendor may assume that each note taken indicates a defect found and the more time taken by the surveyor the worse will be the resulting report.

So the surveyor should explain what he or she is going to do, where necessary asking for permission to proceed. This may seen trivial or obvious but it is surprising in practice how often one hears of surveyors who have been less than courteous, left unrepaired damage or engendered a feeling of hostility in the vendor. Surveyors are representatives of their clients and their profession.

At the end of this book I have listed a glossary of technical terms frequently used by surveyors together with a bibliography of the various publications referred to in the text and a list of other recommended publications by my publishers. There is also a list of the court cases referred to with page references and some useful addresses and web sites for trade bodies and other organisations in the UK to which clients may be referred in some circumstances. Finally I have provided a comprehensive index.

In the present litigious climate the reader might assume that a living which is largely dependent upon Building Surveys is too hazardous to contemplate. However, if one acquires good working knowledge of construction, sticks to facts in the report, ignores hearsay and does not venture too many opinions, then undertaking Building Surveys can make a rewarding career for the surveyor. Every single building is different, each has its own individual atmosphere and character and there is no better way of appreciating the love and craftsmanship which has, over the years, gone into much of our architectural heritage.

All references to legislation or regulations in this book are to those applicable in England and Wales.

1

Organisation

It is highly important to ordinary members of the lay public that a surveyor should use proper care to warn them regarding matters about which they should be warned over the construction or otherwise of a piece of property and that they should be told what are the facts.

Mr. Justice Hilberry – Rona v. Pearce *(1953)*

The purpose of a Building Survey is to give an independent professional opinion on the condition of a property. The traditional relationship is for the surveyor to be acting for a potential purchaser but a surveyor could also be instructed to prepare a report in other circumstances some of which are described later in this book including reports for litigation, dilapidations or for a vendor prior to marketing the property. Various levels of inspection and testing may be appropriate and there are many possible formats for the report.

Site notes, field sheets and checklists

A written record of what was done during the survey is required and should be kept on file for future reference. This may be needed to deal with further enquiries from clients at a later stage and would certainly be necessary in order to defend a complaint or claim. The record should show details of the conditions applicable at the time of inspection, the checks made and what was seen. It should also show, for the avoidance of doubt later on, what may not have been seen and what parts of the building were inaccessible.

The written record should be clear enough that another surveyor could look at the notes later and be able to understand what was done and seen during the inspection. There could be circumstances where the original surveyor is no longer available to translate his or her handwriting and there may be a complaint or claim to deal with.

The normal way to do this is for the surveyor to use an A4 clipboard with pro forma site notes, field sheets and checklists. This will demonstrate methodology. The surveyor will have in front of him or her, the necessary prompts and reminders and a system of recording information. An example of a pro forma field sheet for a Building Survey is shown on the following pages. When completing the form it is always a good idea to consider how it might appear later if it were produced in court in relation to a negligence claim.

So the barrister for the claimant may be cross-examining the unhappy surveyor:

Mr. Glover, did you go into the loft?
Indeed I did and on page 2 of my field sheet there is a sketch of the roof timbers and a check list of the various parts of the roof construction with my comments.

SURVEYOR: **REVISED 5th October 2005**

PROPERTY ADDRESS:

JOB NO.   **ANY ASSOCIATED JOB NO.**

DATE OF INSPECTION:

TIME OF INSPECTION: Start   Finish

TYPE OF INSPECTION: BSR/HBR

WEATHER: Dry/overcast/rain

ORIENTATION: North/South/East/West – Facing

VIEWING ARRANGEMENTS: Keys/agents. Vendor in occupation. Met at property.

LIMITATIONS OF INSPECTION: Furnished   Carpeted   Occupied   Contents

TYPE OF PROPERTY: DH SDH ETH TH SDB DB PBF CF OTHER

DATE OF CONSTRUCTION:

AGE OF EXTENSIONS/CONVERSION ETC:

UNUSUAL FEATURES:

SITE: Level?   Trees?   Subsoil (GSM)?
             Mining area?   Flood plain?

FLATS: Lease term   (years)   Ground rent   (£)   Service charge   (£)
             Common Parts   Lift   Porterage
             Communal heating/hot water   External decorations or other works due?

COMPARABLES (For HBR or BSR with valuation refer to schedule in MVR site notes)

VALUATIONS (For HBR) OMV (£)   INSURANCE (£)
                             Gross External Floor Area m^2
                      (additional notes overleaf . . .)

--1--

ROOF SPACES:

Sketch

Covering

Underfelting

Ridge/Hips

Rafters

Purlins

Struts

Collars

Joists

Hangers

Binders

Wallplates

Ceilings

Insulation

Gable/Party Walls

Wiring

Plumbing/tanks

Ventilation

Gang nail truss: Bracing   Strapping

Limitations on inspection   Action Points

Access hatch   Decking   Insulation?   Underfelting?   Ventilation?

Insulation   Contents   Lag pipes/tanks?   Strengthen timbers?

Ceilings?   Chimney breast support?

Party walls fire stopped? Rot or beetle?
(additional notes overleaf . . .)

FIRST FLOOR:

Sketch

Ceilings (plasterboard, l&p)

Outer walls

Partitions

Floors (firm – joist depth)

Windows

Doors

Radiators (small/micro bore)

Plumbing

Sanitary ware & Bathroom

Wiring

Cupboards

Fittings

Chimneys

Decorations

Limitations on inspection

Floors covered

Furniture

Action points

Plasterwork/ceilings?

Decorations?

Plumbing/sanitary ware?

Windows?

Movement?

(additional notes overleaf . . .)

GROUND FLOOR:
Sketch

Ceilings (plasterboard, l&p)

Outer walls

Partitions

Floors (timber/solid)

Windows

Doors

Radiators (small/micro bore)

Plumbing

Sanitary Ware

Wiring

Cupboards

Fittings

Chimneys

Decorations

KITCHEN (appliances not integrated?)

Limitations of Inspection   Action Points

Floors covered   Plasterwork/ceilings?

Furniture   Decorations?

                                   Windows?

SERVICES: Electricity   Fuses/RCCB unit?   Earth bonding to services?   Any major
               works since 1st January 2005 needs NICEIC or Building Regs.

Gas   Old pipework still live? Flues/ventilation? Meter location?   (*Note*: All boilers
installed after 1st April 2005 **must** be condensing type installed by CORGI with log book)

Water   Supply pipe lead?   Embedded in solid floors?   Water pressure?

Central Heating   Boiler?   Radiators?   Circulation pipework?
                                   (additional notes overleaf . . .)

DAMP, TIMBER, HAZARDOUS MATERIALS

SKETCH

DPC material seen?

Any recent work?

DPC 150 mm clearance to outside levels?

DPC bridged Gutter splash etc?

Timbers

Wood boring beetle?

Wet rot?

Dry rot?

Basement/cellars   Tanking?   Ventilation?   Flood risk?

Limitations to inspections and tests

Action Points

Check with vendor for reports and guarantees?

Specialist report on damp required?

Asbestos Check List
Soffits
Flue pipes
Water tanks
Textured coatings
Corrugated roof sheets
Asbestolux sheets/garage ceilings/boiler cupboards
Pipe and tank lagging
Thermoplastic tiles

Specialist report on timber required?

Lower exterior ground levels?

Attend to rain water goods?

Clear debris from sub floor areas and sweep out?

Improve sub floor ventilation?

(additional notes overleaf . . .)

EXTERIOR ELEVATIONS:

Sketch (Show movement/cracking where applicable)

Chimneys

Flashings/soakers

Roof covering

Parapets/valleys

Eaves

Gutters

Rain water pipes

Walls

DPC

Woodwork

Extensions

Movement

WINDOWS
2000 Building Regs: all habitable rooms except kitchens above ground floor opening windows 0.33 m² min 450 × 450 mm sill max 1100 mm above inside floor
(FENSA/Build. Regs needed for double glazing after April 2002)

Limitations on Inspection

Roof areas not visible?

Ivy/climber on walls?

Flat roofs/parapets not seen?

Adjoining gardens/non-public areas not entered?

Action Points

Woodwork/Redecoration?

Pointing/rendering?

Roof/stacks/rain water goods?

Movement?
Engineers' report?

(additional notes overleaf . . .)

SITE AND DRAINS:
Sketch (Show manholes and drain runs where traced)

TREES

Species

Height/distance

SITE

Outbuildings

Boundaries?

Rights of way?

Shared access?

DRAINS

Drain flow

Trees near drains?

Drains under buildings?

Open gullies?

Limitations on Inspection

Drains not located?

Drains blocked?

Access from curtilage and
adjacent public areas only?

Action Points

Legal adviser, rights of way,
boundaries?

Tree management?

Drains test advisable?
CCTV survey?

Shared drains first used before 1st October 1937 (1st April 1965 Inner London Boroughs) are
public sewers – refer defects to main drainage authority. From those dates shared drains are private
sewers – liability extends beyond curtilage.

(additional notes overleaf . . .)

Equipment

Lighting is needed for roof spaces, under floors, cellars and other dark areas. I find it advisable to carry a lead light with extension lead so that mains powered lighting can be used whenever mains electricity is available. There is better substitute for mains lighting for this purpose and a 100 or 150 W rough service bulb is ideal. When mains electricity is not available one of the modern rechargeable halogen torches is best although the rechargeable batteries to these generally have a fairly short life and they need to be charged up frequently.

Tapes, laser measuring devices and folding wooden rules are useful for taking dimensions. Short steel tapes are best for small dimensions with longer tape measures for external and land measurements. Wooden rules can be inserted between floor boards to check the depth of floor joists.

Ladders are needed for gaining access to roof voids and for inspecting low-level roofs externally. There are a number of 'surveyors'' ladders on the market specifically designed to dismantle or fold up so that they can be carried in a car boot.

An important attribute for any surveyor is that he or she has a good head for heights and is reasonably sure-footed when clambering into lofts or onto roofs. Rubber soled shoes with good grip are advised and you should always take great care setting up a ladder to ensure that it is correctly angled and properly grounded. As a student surveyor I was once sent up a long extension ladder onto the flat roof of a block of flats not, as I discovered later, so that my valued opinions could be given, but to test my resistance to vertigo. Access to roof voids, the roofs themselves and underfloor areas often require strenuous acrobatic feats and as the years go by and the waistline expands these become more challenging.

It would indeed be unfortunate if, many years from now, the skeleton of a surveyor should be dis-covered, firmly wedged between the rafters in an empty and long abandoned building. They would know that it was a surveyor because the clipboard and notes would still be there with a final note 'rafters closer together than normal spacing'. So it is a good idea to make sure that someone back at the office knows where you are when working alone, and that you carry a mobile phone.

Moisture meters are necessary for testing plaster and woodwork for damp and the instrument should be tested prior to use on each occasion since it is important that a meter with failing or flat batteries should not be used since its readings will then be inaccurance and unreliable.

Binoculars will provide a clearer view or roofs and upper storeys and a selection of tools and manhole-lifting keys are required for gaining access to the drains. A hammer and bolster are useful for easing tight manhole covers. A long spirit level is needed to check plumbness in walls and whether floors are level. A plumb line could be useful for checking the verticality of walls from upper windows or balconies. A selection of tools including probes to check timber for rot and a key for opening electric and gas meter boxes should be carried. Mirrors are useful for peering around corners under floors and within ducting.

For general purposes I carry the following equipment in my car:

1. A4 clipboard with pro forma site notes, field sheets and checklists.
2. Folding or sectional 'surveyors'' ladder.
3. Laser measuring device, linen and steel tapes and wooden rule.
4. Mains powered light and extension lead plus rechargeable torch.
5. Moisture meters.
6. Binoculars.
7. Spirit level and plumb line.
8. Selection of manhole keys, claw hammer and bolster.
9. Mirrors.
10. Identity documents.
11. Selection of tools including meter box keys, screwdrivers and wood probes.
12. Protective gloves and face masks.
13. Digital camera.

Fig. 1.1 Equipment for a Building Survey of a simple low rise structure.

14. Magnetic compass.
15. Endoscope.
16. Drain plugs and drain testing equipment.
17. Metal detector.

For more specialised inspections other equipment are available; hand augers for taking soil samples and shear vane testers for testing clay soils, for example, or concrete core sampling equipment; calibrated tell tales may be used for monitoring crack damage in walls; radon gas detectors may be used. However, none of this specialised equipment is likely to be used during the course of the initial Building Survey.

Instructions

All instructions should be agreed in writing and the clients should be asked to sign agreed terms and conditions of engagement. The surveyor's instructions may be limited to inspecting a particular building or part of a building and providing a report without any need to comment upon the price or related matters so the report will be purely concerned with the construction and condition of that property. Such reports used to be called 'structural surveys' although the generally accepted terminology nowadays is 'building surveys' because the use of the word 'structural' suggests some kind of intrusive investigation into the structure of the building which is not normally practical with an initial building survey report (although more detailed intrusive investigations could well follow).

In some cases the report may also include advice on value when it might be termed a 'building survey and valuation'. Often a building survey is being undertaken in conjunction with a valuation for a mortgage lender for a combined fee. Building surveys can be undertaken on flats, maisonettes or other units of accommodation which form part of a larger building in which case it is clearly vital to confirm at the

Fig. 1.2 Borescope for inspecting inside wall cavities, sub-floor voids and other areas. The wide-angle lens provides good depth of field.

Fig. 1.3 Environmental issues should be covered in the report but the client must decide whether a problem such as aircraft noise is acceptable or not.

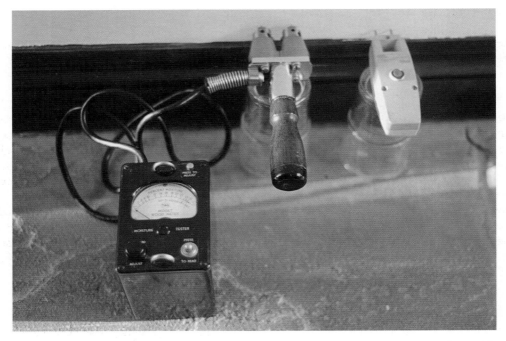

Fig. 1.4 Moisture meters are not a new invention. On the left a Weston Moisture Meter manufactured by the Weston Electrical Instrument Corp in the USA in the 1930s. Principally designed to check moisture levels in timber but also having a scale to measure dampness in plaster. On the right an Aquatrace made in England in the 1960s – you press the red button and a flashing light indicates damp. The faster it flashes the damper the wall.

outset – in the terms and conditions of engagement – precisely what is being inspected and the limitations of the report in respect, for example, of the interior of other flats in the same building.

Fees

The fee should be collected in advance. Nowadays many clients pay the fee by telephone using a credit card or instruct the surveyor on line using a web site. It is still necessary to have signed terms and conditions of engagement, however, before proceeding with the inspection. There are no fixed scales of charges for building surveys so each firm of surveyors is able to devise a scale based upon a suitable formula.

When dealing with a high volume of survey instructions it is probably easiest to adopt a scale of charges based upon the value of the property, although this can be unfair because the amount of work involved in surveying a small modern house in a high-value area will clearly be much less than a rambling old house in a low-value area with the potential liability commensurately higher in the latter case. So the scale of charges might also be weighted to allow for the age of the property amongst other things.

If the survey is being undertaken in conjunction with a mortgage valuation for a prospective lender then a single fee for both services is normally agreed providing some saving for the client.

The office

Invariably nowadays the office organisation will primarily be undertaken using a computer system with suitable survey management software and the details of terms and conditions of engagement with a copy of the report itself and any subsequent correspondence will be stored there. However, a paper filing system will also be needed to hold site notes, field sheets, checklists, photographs and other information gathered in the course of preparing the report.

Having regard to the possibility of litigation arising, perhaps many years after the event, it is advisable to keep the old paper files in secure storage for a long time and certainly for at least 15 years.

Standard paragraphs and phrases

In Chapter 18, I have included a typical building survey report and this report includes a number of paragraphs which could be stored in the word processing software as standard paragraphs or phrases so that when the report is dictated the surveyor need only refer to the code for a particular paragraph and it is then automatically included in the draft. Of course, surveyors have always used standard phrases but they were not always stored in the software; surveyors had them in their heads and when the reports were dictated the standardised text just came automatically and instinctively.

However, we need to be careful about excessive use of standard text. I remember hearing about a client of one of the large national firms of surveyors who rang up the office having received his report and complained that since the report had obviously been written by a computer he should have some discount from the fee because he was not getting the individual and undivided attention of the surveyor which he thought he was getting when he gave instructions.

Clients

Finally, I come to the clients without whom, of course, surveyors would go hungry. The relationship between the surveyor and the client is a contractual relationship whereby a service is provided in exchange for payment and the basis of that service is governed by the terms and conditions of engagement signed and agreed by both parties. Notwithstanding disclaimers to the contrary the surveyor might have a liability also to third parties and the terms and conditions of engagement may themselves be subject to a test of reasonableness under the Unfair Contract Terms Act. The question of liability is dealt with in more detail in Chapter 19.

Ideally the surveyor should discuss the client's particular requirements before inspecting but this will often not be possible in a large surveying practice where trained secretarial staff take the instructions and the surveyors themselves are away from the office much of the time stuck in lofts or with their heads down manholes. So the staff taking instructions need to establish if the clients have special concerns or requirements in respect of the report.

Sometimes clients will be buying a property with a view to a change of use, making alterations or building extensions. Various consents may be needed including, possibly, planning permission, listed building or conservation area consent, approval under building regulations or approval under the terms of restrictive covenants. In the case of leasehold property the freeholder's consent will invariably be needed. If permission to undertake works or change the use is material to the viability of the purchase then clients should be advised to postpone any legally binding contract until the outcome of applications is known, or enter into a conditional contract.

Some disabled clients may have difficulty inspecting all parts of the property themselves and could have special needs in respect of wheelchair access so it is helpful if the survey is aware of this prior to inspecting. I have prepared reports for blind clients who would have the report read to them and for whom a number of potential hazards around the house would need to be identified.

Clients buying commercial buildings will have particular requirements in respect of services; three phase electricity or suspended floors for computer cabling, for example, or loading bay headroom and vehicular access, so all this needs to be established at the outset prior to the survey.

So there we have it. The office is organised, the surveyor has the necessary computer system to compile the report, the equipment is in the boot of the car, the client's requirements have been established and the fee has been paid. All that remains is to carry out the Building Survey.

2

House surveys

We do not need a Building Survey – if you can just have a quick look at the house to make sure that there are no defects to the building, that will be fine.

Extract from client's letter of instructions

Inspection procedures

You should follow a methodical procedure which will often be determined by the layout of the site notes or field sheets being used. There is no right way or wrong way to inspect a house but in my view it is generally preferable to begin on the inside at the top and finish on the outside at the bottom, if for no other reason than that you are less likely to subject the owner's carpets to wet or muddy feet and ladders. For example, the procedure for a conventional dwelling house might be:

1. A preliminary inspection of the whole property to establish the type and layout and to determine what ladders and other equipment may be needed.
2. An inspection of the accessible roof spaces.
3. A room-by-room inspection at each floor level starting with the topmost floor.
4. An inspection of accessible basements, cellars and sub-floor areas.
5. An examination of the roofs externally using binoculars and surveyors' ladder.
6. An examination of the various elevations.
7. Inspection of site boundaries, outbuildings and surroundings.
8. Location and inspection of the drains.

Detailed procedures

The recommended detailed procedures under each of the preceding headings are as follows.

Preliminaries

The surveyor notes the general character and description of the property, that is, whether detached, semi-detached, terraced, end of terrace and number of storeys. If terraced it is advisable to note the overall length of the terrace and check for obvious signs of movement or rebuilding to other parts of the terrace.

The foundations serve a detached structure in isolation and that structure will not normally require provision for movement unless it is very large. A long terrace or a large block of flats has to be viewed as a single structure and the surveyor must consider if the entire building has been designed to resist the forces to which it will be subject including possible differential ground movements and thermal expansion and contraction.

Many examples of terraces of old house will be encountered, some very long indeed, built with no provision for movement. It is always a good idea to look at the flank walls at each end of a terrace for signs of leaning or bulging.

The surveyor must note the general lie of the land and the gradient on which the building stands. Houses built on sloping sites are obviously more likely to suffer from structural movements as compared with houses built on the level and in the subsequent inspection the surveyor will be looking from signs of past structural movement (see *Morgan* v. *Perry* p. 186).

Reference should be made to the Geological Survey Map for the area prior to inspection and the surveyor will be placed on guard against the possibility of problems if the subsoil shown is a shrinkable clay, or peat, or if old mine workings, gravel pits or quarries are shown. In areas affected by Radon gas, the Radon map will have been checked and in an area affected by past coal or other, mining records of this will be consulted.

Geological Survey Maps, solid and drift editions, are available from British Geological Survey Information Office, Natural History Museum Earth Galleries, Exhibition Road, London SW7 2DE. The Radon Atlas of England and Wales is published by the National Radiological Protection Board, Chilton, Didcot, Oxon OX11 0RQ. Coal Mining Searches Directory and Guidance produced by the Coal Authority and the Law Society is published by the Law Society, 113 Chancery Lane, London WC2A 1PL.

Perhaps, the most important matter to be established at the outset is the age of the house because the age will inform much of what the surveyor will be looking for during the detailed inspection. The defects likely to be encountered in a house built in 1950 will be very different from those in a house built in 1850.

Fig. 2.1 This house is built directly on the riverbank. Issues arising include foundations, potential riparian liabilities and flood risk.

In the 1950s house we might expect to find cavity wall tie failure, asbestos–cement materials, wet rot to external woodwork, poor quality flat roofs and other problems associated with that era but not the wood-worm, wet and dry rot, defective lath and plaster and roof problems common in Victorian construction.

Roof spaces

A sketch is a good idea, showing the layout of roof timbers with a note of the principal timber sizes and spacings. The type of covering, whether underfelted or not should be noted together with the condition of any gable or party walls, valleys and ceilings. Depth and type of ceiling insulation should be noted and any accessible wiring and plumbing checked. Detailing around chimneys should be inspected and if any chimney breasts below have been removed the manner of support for oversailing brickwork or stonework should be confirmed.

A lot can be learned about the history of the building within the roof space and in most old houses the main defects are likely to be found either at the top or at the bottom. It goes without saying that great care should be exercised when moving about within the roof. If there is a double layer of insulation the location of ceiling joists or their suitability to take the weight of a well-nourished surveyor may be doubtful. The surveyor is not expected to take undue risks on behalf of his or her, client.

Room-by-room inspection

You have to be methodical about this and avoid distractions. The writer draws a sketch and marks on this type of ceilings, walls, floors, windows and fittings. Power and lighting points are marked together with radiators. Clients will want to know if there are very few, or no, power sockets or if a particular room lacks a radiator.

Surveyors normally tap the walls and ceilings to check what they are made of, and stamp on the floors for similar reasons. With experience you can distinguish between ceilings and stud walls surfaced in plasterboard from those surfaced in lath and plaster. Similarly, the difference between a suspended timber floor and a solid concrete floor is usually fairly obvious. However, there will be cases where the result of this type of stamping and tapping is inconclusive and if you cannot confirm what a ceiling, wall or floor is made of, then say so in the report and follow this up with any necessary additional advice.

It is at this stage that any damp testing is undertaken and a careful check made around the skirtings, under windowsills and at other vulnerable points for damp and associated timber defects. Doors should be opened and closed and door heads checked for slopes indicating past movement.

Basements, cellars and sub-floor areas

As with the loft a great deal of useful information about a house can be gleaned from a cellar so if you are fortunate to find a cellar or basement spend some time there checking the accessible construction and services. Take care when descending the cellar steps since rot and beetle attack to timbers in cellars is very common and sometimes the steps are unsafe.

Cellars below damp-proof course, and not tanked, will be inherently damp and need good ventilation. Check for signs of flooding since if the storm drainage system around the house becomes overloaded during heavy rainfall the cellar may actually take in water. It is not uncommon to find a sump set into a cellar floor with electric pump operating from a ball valve, to pump out ground water percolating in through the sides, if the cellar is below the surrounding water table.

External inspection of roofs

The number of chimney flues should be noted so that a check can be made to trace these through to the fireplaces below. If fireplaces have been removed and the flues are redundant then air-vents should be provided at the base of the flues with hooded pots or ventilated caps at the top to provide a flow of air up the flue to keep it dry. If there are open fires or appliances using the flues then clients should be advised to have the condition of the flue linings confirmed and a smoke test carried out.

The condition of parapets, flashings soakers and other detail should be noted. Tile dentils or cement fillets are often used to seal the joint between tiles, slates or other coverings and the adjacent stacks and walls. These are rarely very satisfactory in the long run and often crack and leak. Properly formed lead flashings using code five lead or similar, well chased and pointed into the brickwork are much to be preferred.

The sloping roof coverings should be examined paying particular attention to cracked, slipped or loose tiles or slates. Hip and ridge tile bedding mortar will often be found to be cracked and loose. Generally, this should not be repointed but the tiles lifted and rebedded in new mortar using a suitable hip hook to restrain the bottom tile.

The roofs over older terraced buildings, especially Victorian and Edwardian houses, often feature fire walls in the form of sloping parapets carried up on top of the party walls and the detailing to these is frequently poor. Good flashings are needed to seal the joints with the tiles or slates. Modern practice is to provide flashings only dressed down over the tiles in contrast to the traditional (and better) flashing and soaker arrangement. Walls should be finished with throated copings stones set on damp-proof courses and the pointing to brickwork or stonework needs to be in good order. Beware of vegetation such as weeds or even small saplings found growing out of the brickwork to this type of parapet – a sure sign of excessive weather penetration and possible problems with plasterwork and timbers below.

Where accessible flat roofs should be inspected from windows above or using a surveyors' ladder. If you go onto a flat roof take care since there is a possibility of soft or rotten decking which could be in danger of collapse.

The elevations

I recommend that each elevation be sketched using a simple line drawings. If cracks are present they can be shown on the sketch with the crack size indicated using suitable notation. I have found that photography is often not much use with the smaller cracks which do not show up well in a photograph although is large significant fractures are present then photographs of these would be worthwhile.

Window and door openings should be checked to see if they are square and walls generally to see if they are plumb. Movements due to spreading arches, sagging bay bressumers and settlements should be noted. Look out for leaking down pipes and gutter spash at the base of solid walls which may well correspond to areas of dampness found inside.

The site and surroundings

The condition of boundary walls and fences should be noted and especially any retaining walls. Rebuilding boundary or retaining walls can be very expensive should such be required. Driveways and other paved areas should be checked. Cracked and uneven paving could raise issues of liability to visitors. Paving should always be at least 150 mm (two brick courses) below the damp-proof course or internal floor level whichever is the lower, with a slope away from the walls to good storm water drainage unless some form of basement tanking or vertical damp proof coursing is present.

Outbuildings, swimming pools and other facilities should be inspected to whatever standard is required by the signed and agreed terms and conditions of engagement. Frequently these may be excluded from the report.

Drains

Hopper heads, soil vent pipes and other drainage above ground should be inspected and accessible manhole covers within the curtilage lifted in order to check the drainage system below ground. Having regard to the risk of damage and the health and safety issues arising the surveyor is not expected to remove manhole

Fig. 2.2 The huge oak tree almost touches the garage wing to this house and there are a pylon and overhead power lines close by, both matters to be reflected in the report.

Fig. 2.3 Artificial stone cladding generally reduces the value of the property to which it is applied, and in this case by cladding only at ground floor level the owner has left a ledge for driving rain to penetrate behind the cladding resulting in a damp problem inside.

covers, which are tightly rusted to their frames or to lift heavy duty covers which cannot be conveniently removed by one person.

The surveyor needs to be aware of the local water regulations which apply to drainage. For example in inner London, thanks to our Victorian forebears, we have an old combined system which takes all foul water and storm water to the same sewer. Unfortunately the system can become overloaded in heavy rainstorms leading to flooding in basements and cellars. In outer London and most other parts of the country we have separate storm water and foul water sewers and it is against local regulations to run storm water into the foul water drain. When a storm sewer is not available the storm water will typically be run into soakaways in the form of land drains or soakaway pits.

In relation to liability for the drains the surveyor needs to record on site if the drains are shared and estimate the date when they first came into use. In England and Wales a shared drain which first came into use prior to the 1st October 1937 (1st April 1965 Inner London Boroughs) is a public sewer and the responsibility of the local main drainage authority. This applies notwithstanding that the shared drain may run through private gardens before connecting to the main sewer in the highway. So if it blocks or needs repair the main drainage authority is responsible. Drainage authorities tend not to publicise this too much and as a consequence many householders pay for drains to be unblocked or repaired when they could have had the work done free by telephoning the local office of the main drainage authority.

In my opinion a surveyor should not undertake any testing of drains during the initial Building Survey inspection. There are a number of reasons for this. The drains may be old and in poor order so that a water pressure test could make matters worse. There may be insufficient access to enable drains testing to be undertaken. But there are a number of issues affecting the drains which can be covered in the report based just on a visual check within the manholes.

Fig. 2.4 If leaning boundary walls have to be rebuilt this can be very expensive.

The manhole covers should be close fittings and airtight. Drains should be flowing freely with no signs of blockage and the manholes clean. There should be no tree roots entering manholes. Cement benching at the base of manholes and the channels themselves should be free from cracking or erosion. Vent pipes should terminate at least 900 mm above any adjacent opening window. The old type of open gulley where the waste pipes discharge over a grating is no longer regarded as hygienic and replacement with a modern back-inlet gulley (where the pipes discharge below the grating) should be advised.

It is worth mentioning here that surveyors dealing with inspections and tests of drains should wear gloves and take precautions against infection, especially if they have open cuts or abrasions. A periodic tetanus injection is advised in order to maintain immunity. Facilities for washing should be arranged after dealing with drains, drainplugs, inflatable bags and other equipment.

Dealing with occupiers

Ideally, the survey should be undertaken when the property is unfurnished and unoccupied and especially if the floor coverings have been removed. However, this will often not be the case and the surveyor will have to inspect a residential property when it is occupied and furnished with fitted floor coverings including – increasingly – the ubiquitous laminated wood flooring.

The surveyor in these circumstances has no special rights other than to look, tap on walls and stamp on floors. Great care should be taken dealing with the occupier's furniture and floor coverings bearing in mind that the surveyor in this context is acting as servant of the client so that the client could be liable ultimately in law for any damage caused. Few clients would take kindly to receiving a bill for damage caused by their surveyor during the course of his or her inspection.

Anyone for trellis?

During the course of a residential inspection it is advisable to look fairly carefully at the boundaries and consider whether the actual extent of the garden and other land included with the property is obvious and if there are rights of way in respect of private access roads, shared drives or footpaths.

Any reader interested to discover the full horror of a litigated boundary dispute would be advised to read the law reports of *Scammell and others* v. *Dicker* (2005) EWCA Civ 405 and (2005) 42 EG 236. In this case, which had been going on since 1989 and which found itself before the Court of Appeal in 2005 the appellant Mrs Dicker was seeking a declaration as to the exact line of the boundary between her property, Hillside and an adjoining property Glenwood Farm, near Wimborne in Dorset. The history of the case reveals a sorry tale of intransigence and serious professional incompetence and as a consequence after 15 years in the courts everyone seems to have lost.

In the end, the respondent's litigation was being funded by the Solicitors Indemnity Fund, due to failures on the part of their solicitor, and the appellant, who had been reduced to poverty by the litigation, was receiving legal aid.

If there is a title plan then the ownership of boundary walls and fences may be shown by a mark in the shape of a 'T' against the boundary line on the side of the owner and this may also be referred to in the title covenants. Alternatively, the boundary may be a party wall or a party fence jointly owned and jointly maintained. Frequently there is no indication at all in the title deeds of who is responsible for a boundary.

There are certain well-established common law rules and customs regarding boundary ownership. So in the case of close boarded fences the 'nails are driven towards home' so the posts and arris rails are on the owner's side and the boundary is the outer face of the boarding. If there is a ditch and bank with a hedge on the bank the boundary will be the centre line of the hedge and the owner will be assumed (in the absence of evidence to the contrary) to be the person who owns the land containing the ditch, since it is expected that a farmer would excavate a ditch on his own land to build a bank on the boundary.

Accurate setting out of boundaries and accuracy in the corresponding title plan is important. In some continental countries – notably Switzerland – permanent boundary markers are set into the land at the corners of plots following a survey by government land surveyors. Here in the UK we rely on the large-scale ordnance survey maps as a basis for title registration. 1:1250 maps may not be accurate enough, however, to resolve disputes about small differences in boundary line.

One hundred typical defects in residential property

This is a list of the one hundred most common issues raised in my experience in Building Survey reports requiring action by the client or giving rise to some warning in relation to a future repairing liability or affecting the value of the property. This list is not intended to be exhaustive but should be of interest.

1. Materials containing asbestos.
2. Cracks to walls.
3. Condensation.
4. Poor quality flat roofs.
5. Rot to external joinery especially lower frames and sills.
6. High external ground levels bridging damp-proof courses.
7. Leakage from poorly designed shower enclosures.
8. Poor sound proofing in converted flats.
9. Sub-standard electrical installations.
10. Defective drains.
11. Non-compliance of gas installations.
12. Lead plumbing.
13. Contaminated land issues raised by legal advisers following environmental searches.
14. Defective heating systems.
15. Bulging flank walls.
16. Wall tie failure.
17. Condensation in double-glazed sealed units.
18. Spalling and frost damaged brickwork and stonework.
19. Sub-standard loft conversions.
20. Damp basements and cellars.
21. Inadequate bay window support over double glazing.
22. Old roof coverings lacking underfelting.
23. Flooding in the locality.
24. Radon gas.
25. Mining subsidence in locality.
26. Mundic construction in Cornwall.
27. Settlement of solid ground floor slabs.
28. Woodworm, mostly Common Furniture Beetle.
29. Dry rot under floors, behind panelling and in flat roofs.
30. Leaking parapet and valley gutters.
31. Inadequate means of escape in the event of fire from flats.
32. Poorly insulated ceilings in lofts.
33. Lack of party/fire walls in old terraced roof spaces.
34. Sagging and spreading of roof timbers.
35. Condensation in old chimney flues.
36. Failure of damp-proof courses.

37. Leakage of water into wall cavities from concrete 'Finlock' gutters.
38. Cracked and leaning boundary and retaining walls.
39. Tree roots posing risk to structure.
40. Coastal erosion.
41. Failure of old lath and plaster ceilings and partitions.
42. Leaking gutters and down pipes.
43. Inadequate support when chimney breasts are removed.
44. Shortage of power sockets.
45. Inadequate ventilation to bathrooms and showers.
46. Inadequate ventilation to kitchens.
47. Poor septic tank and cess pit drainage arrangements.
48. Inadequate sub-floor ventilation to timber ground floors.
49. Excessive service charges to blocks of flats.
50. Cracked and loose external rendering.
51. Poor woodwork and sash cords to sliding sash windows.
52. Hazardous large sheets of ordinary quality glass to walls and partitions.
53. Excessively steep staircases and inadequate balustrades.
54. Flat roofs used as terraces without proper wearing surface.
55. External electrical circuits not in armoured conduit.
56. Adjacent high-voltage pylons, electrical sub-stations and mobile phone masts.
57. Problems with shared drives and private access roads.
58. Inadequate management arrangements for blocks of flats.
59. Damp and rot to timbers due to leaking washing machines and dishwashers.
60. Cracking and shelling to wall plaster.
61. Lamination of clay and concrete roof tiles.
62. Nail sickness and cracking to old slate roofs.
63. Lack of wind bracing to modern gang nailed truss rafter roofs.
64. Damp penetration and staining to walls from leaking overflow pipes.
65. Settlement of blockwork partitions resting on timber floors.
66. Poor quality supporting platforms for water tanks and cylinders.
67. Excessive bounce to floors due to over spanned joists.
68. Leaning chimney stacks.
69. Leaking service and heating pipework embedded in solid floors.
70. Sub-standard extensions converted into habitable rooms.
71. Inadequate covers and lagging to water tanks in lofts and on flat roofs.
72. Rot to floors under leaking flush pipes and waste connections to water closets.
73. Builders' rubble and other debris left under floor boards blocking ventilation.
74. Galvanic corrosion to old galvanised steel water tanks.
75. Header tank overflow pipes run into adjacent cold-water tanks.
76. Foundation settlements on shrinkable clay soils.
77. Foundation settlements due to leaking drains causing subsoil erosion.
78. Lifting of foundations by ground heave due to clay swell where trees removed.
79. Inadequate damp proof membranes in solid concrete ground floors.
80. Integral garages with inadequate protection against spread of fire into the main building.
81. Wet rot to painted softwood frames to porches and conservatories.
82. Cavity walls lacking damp proof cavity trays and other detail.
83. Tree roots in drains.
84. Inadequate encasement of steelwork and insufficient cover to steel reinforcement.

85. Removal of internal walls and chimney breasts compromising structural stability.
86. Use of spiral and open riser staircases which do not comply with Building Regulations.
87. Porous thatched roofs attacked by squirrels and other vermin.
88. Brick and blockwork garages with inadequate piers and/or roof strapping.
89. Poorly installed and potentially dangerous loft ladders.
90. Inadequate cavity fire checks, breather membranes and other detail in timber frame houses.
91. House Longhorn Beetle in the Weybridge area of Surrey.
92. Lack of asbestos management arrangements for common parts to blocks of flats.
93. No proper pavement crossover for forecourt parking.
94. Storm water soakaway systems completely blocked and needing relaying.
95. Hot-water pumping over into loft header tanks.
96. Furring up of cylinders and pipework in hard water areas.
97. Failure to obtain listed building or conservation area consent where these apply.
98. Thermal expansion and contraction causing cracks to Calcium Silicate brickwork.
99. Death watch beetle in old damp hardwoods especially Oak.
100. Neighbours undertaking works to party walls without necessary agreement.

It is interesting to compare this list with the list I prepared in 1980 for the first draft of this book and how clients' concerns have changed. The three main issues raised by clients still seem to be cracks, damp and the roof. But we now also have asbestos, radon, flooding, contaminated land and electromagnetic fields generated by pylons and mobile phone masts together with a much greater concern with the standard of finish and how the services work. Some of these issues have arisen as a consequence of the additional environmental and other searches made by conveyancers and media coverage.

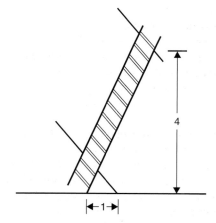

Fig. 2.5 During the survey any ladders used must be properly angled and grounded. A 4 to 1 ratio of height to distance is best. Steeper and the user may topple backwards; shallower and the ladder may slide from beneath.

3
Foundations

And every one that heareth these sayings of mine, and doeth them not, shall be likened unto a foolish man, which built his house upon sand; and the rain descended, and the floods came, and the winds blew and beat upon that house; and it fell: and great was the fall of it.

St. Matthew VII 24–27

I well remember the long hot summer of 1976. The previous year had also been quite dry so we started with reduced water reserves in the reservoirs, and as 1976 progressed the dry landscape of England came increasingly to resemble that of Spain or even North Africa. Properties built on clay soils began to crack and by the autumn there was widespread subsidence damage to buildings especially in the London area.

Readers familiar with the workings of insurance companies will know that they rarely give away something for nothing but a few years earlier, at the beginning of the 1970s, the insurance companies, at the behest of the building societies and other lenders, had agreed to extend the cover offered in the standard household insurance policies they issued. These policies, which started life just covering properties for fire, had been progressively expanded to deal with storm damage, burst pipes and other perils. And in the 1970s the cover was extended to include subsidence and landslip (and later heave) with the general level of premium unchanged.

An unintended consequence of this largess by the insurance companies was that following the drought of 1976 there was a massive increase in the number of insurance claims for subsidence and many builders, seeing an obvious business opportunity, set themselves up as underpinning contractors and started to tout for the repair work which was being funded by insurance claims. Claims escalated and the unhappy insurers found themselves paying out hundreds of millions of pounds in exchange for no commensurate increase in premiums.

When undertaking a building survey, it is essential that the surveyor has a full understanding of the principles involved in foundation design both in terms of modern practice and also what was done in the past, and it is important to be able to distinguish between evidence of movement and cracks which may be of little consequence (although perhaps alarming to the client), and those which may indicate something seriously wrong below ground. In those cases where further investigation is needed then it is normal to refer the client on to a structural engineer unless the surveyor concerned is also qualified in that field.

Design

Modern foundations for simple low-rise buildings are likely to be either shallow strip, deep strip, pad-and-beam, pile-and-beam or raft. Each type has advantages and disadvantages and will be suitable for some conditions and unsuitable for others. It should be bourne in mind that builders will generally use the most economical foundation which will satisfy the building inspector or district surveyor at the time, and fashions in foundation design change.

Older buildings erected before modern concretes became generally available and often had very little in the way of foundation at all. Common practice in the 19th century and earlier, especially in rural areas, was to dig shallow trenches and lay a brick footing on a bed of hoggin perhaps about 500 mm below ground level. It is surprising how well many buildings erected in this way have stood the rest of time.

Surveyors often find themselves inspecting foundations exposed for some reason, and it is common to find that the actual foundations differ from what the plans, or normal practice, would have required. Often the foundation depth in particular will be shallower than expected, indicating poor supervision during the course of the works because the builders excavating the trenches would have stopped digging as soon as they could persuade whoever was in charge that they had dug far enough.

I remember once supervising the construction of a small shop extension in Staines in Middlesex close to the Thames where the subsoil is a very variable mix of alluvium and river terrace gravels. The builders had to dig the trenches by hand because there was no access for machinery. Every so often they would stop and ask me to look at the bottom of the trench. At one point they complained that they must, by now, be down to Roman Staines if not deeper and surely they could now stop and place the concrete.

I was not happy and told them to carry on digging. Shortly, after they uncovered a fluorescent light tube. Not something one associates with the 1st century BC.

Foundations are not normally exposed for inspection during a building survey. The Royal Institution of Chartered Surveyors (RICS) practice note does not require this and most standard terms and conditions of engagement specifically, exclude this type of intrusive investigation in the first instance.

A surveyor must not be afraid of stating an opinion on a foundation when sure of his or her ground (in both senses), neither should surveyors be reluctant to refer a client to a specialist engineer when this seems appropriate. This is when local knowledge can be invaluable. Some surveyors have actually seen the buildings being constructed in their area and know first hand what type and depth of foundation was used.

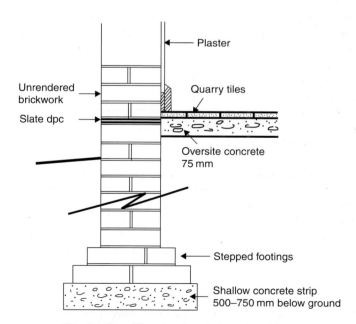

Fig. 3.1 Late Victorian foundation, circa 1900.

The old Building Regulations laid down the following standards for foundations:

> *Foundations must (a) safely transfer all dead, imposed and wind loads to the ground without settlement or other movement which would impair the stability of, or cause damage to, the building or any adjoining building or works; (b) be taken down below frost damage or soil movement levels; be resistant to attacks by sulphates or other deleterious matters in the subsoil.*

In England and Wales, a depth of 450 mm is sufficient to place foundations below the level of frost damage in the most extreme conditions for which provision ought reasonably to be made. In practice a greater depth than this is generally advised because ground conditions on the surface of a site may vary, and foundations must always be taken down below the level of any disturbed topsoil. A minimum depth of 900 mm is therefore recommended unless the bearing subsoil is a hard rock formation or similar base of undisputed strength and regularity.

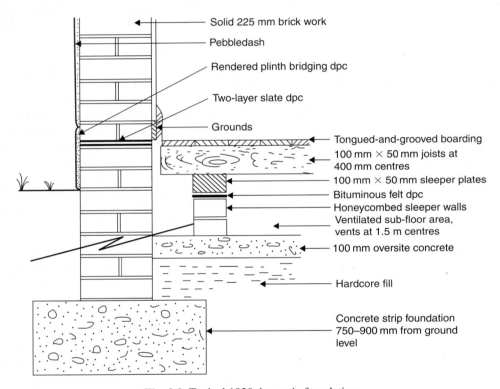

Fig. 3.2 Typical 1930 domestic foundation.

Subsoils

Care is required on chalk soils to ensure that the foundations are at least below the minimum for frost damage since chalks can be subject to considerable frost heave close to the surface.

Despite Biblical claims to the contrary, sandy soils make good foundation bases in many cases, although high-frequency vibrations in the range of 500–2500 impulses per minute can cause serious settlement problems, a point unlikely to arise with residential structures unless near to machinery. This factor should be borne in mind when dealing with industrial structures, particularly in the case of changes of use involving machinery.

On shrinkable clay soils a foundation depth of 1200 mm is recommended which would be below the level of seasonal change on level sites in the absence of vegetation. On sloping sites rather greater depths are required, especially on south-facing slopes where the ground will drain and dry out readily during periods of dry weather.

The presence of trees or bushes on shrinkable clay sites causes a particular problem for surveyors advising a prospective purchaser. The standard against which the property should be judged should be that set out in the appropriate British Standard. (BS 5837, 1980 'Code of Practice for Trees in Relation to Construction'). This BS states: 'Where the presence of shrinkable clay has been established and where roots have been found or are anticipated in the future, it is advisable to take precautionary measures in the design of the building's foundation (in accordance with CP 2004, 1972 and CP 101, 1972). Alternatively, consideration should be given to designing the superstructure to accommodate any foundation movement that might be induced by clay shrinkage. Where a subsoil investigation has not been carried out, an approximate rule-of-thumb guide is that on shrinkable clay, if risk of damage is to be minimised, special precautionary measures with foundations or superstructure should be considered where the height or anticipated height of the tree exceeds its distance from the building.'

One should pause for a moment to consider the full implications of BS 5837. In most cases the property being inspected, whether residential, commercial or industrial in character, is unlikely to have special foundations of the type anticipated, unless it is quite modern. Generally a short-bored pile-and-beam foundation, or pad-and-beam, or occasionally a raft design, would be used to accommodate the seasonal ground movements induced on clay sites by tree roots. (A sketch section through a pile-and-beam foundation is shown in Figure 3.3.) It follows that for most older structures on clay sites a traditional shallow foundation design may be assumed, and that if trees are present whose height at maturity will exceed their distance from the walls of the structure, then the structure is at risk from shrinkage of the clay subsoil and damaging foundation movements would be possible during a dry summer.

Accordingly, the surveyor should make a note of the locations of all trees and bushes near the building being inspected and record their distances from the main walls, and identify the species of tree involved, confirming whether or not the subsoil may be shrinkable. It should also be confirmed whether specific trees or bushes may be within the curtilage of the property under review, or in adjoining curtilages.

The surveyor should then comment and warn the client as necessary in the light of current knowledge. Three separate situations may arise. A building with shallow foundations on shrinkable clay may have trees within the curtilage closer to the structure than their mature height. In similar circumstances there may be such trees but they may be located in an adjoining curtilage under other ownership. Possibly there may be trees within the curtilage of the subject property which, while well back from the property concerned, could be a threat to an adjoining structure and thus give rise to a legal liability in the future.

If trees are closer to the building than the recommended distances, and within the curtilage, some action in the nature of removal, crowning, pollarding or simply subjecting to a period of observation and future pruning, may be advisable depending upon the circumstances.

If trees in other ownership are closer than would be recommended then the surveyor could advise the client to write to the owners of these trees, pointing out that their root systems could be hazardous to the property he is acquiring, and suggesting that further advice be taken in the light of the fact that by virtue of the laws of nuisance the owner of a tree is liable in law for damage caused by its roots extracting moisture from the subsoil under adjoining structures. (See *Bunclark* v. *Hertfordshire County Council* (1977) 243 EG 455 reported in *Estates Gazette*, 30 July 1977.)

Conversely, in a situation where trees within the curtilage of the property being acquired could pose a threat to adjoining structures, then the client should be warned of his own potential liability under the law of nuisance in case this should become significant in future years.

Certain species of tree are really unsuitable in a garden setting on shrinkable clay soil, particularly poplars, willows and elm, and if these are found they should be regarded with a critical eye. Poplar roots

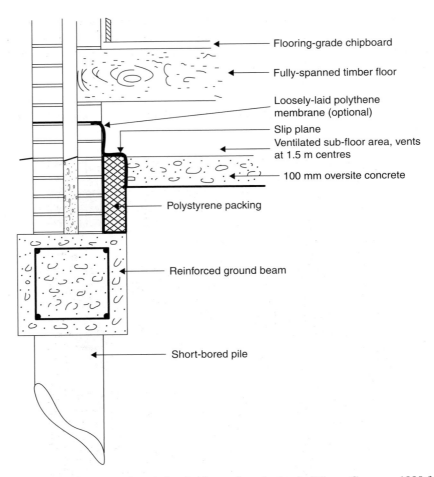

Flooring-grade chipboard

Fully-spanned timber floor

Loosely-laid polythene membrane (optional)

Slip plane
Ventilated sub-floor area, vents at 1.5 m centres

100 mm oversite concrete

Polystyrene packing

Reinforced ground beam

Short-bored pile

Fig. 3.3 Modern short-bored pile foundation for domestic and other buildings. Common 1990 foundation.

can extend for a distance of up to twice the height of the tree in search of moisture. I have also found ash trees to have searching root systems, and a particular liking for entering cracked drains and disturbing paving, quite apart from their possible effects on shallow foundations.

It is recommended that surveyors practising in localities where shrinkage clays are found should familiarise themselves with the various aspects of this particular problem, and that they should be able to identify any species of tree encountered during their work where this seems necessary. The reader may find *Trees in Britain, Europe and North America* by Roger Phillips (Pan Books) to be a most useful guide to tree identification throughout the seasons.

One factor which can commonly cause problems with older foundations is subsoil erosion caused by cracked or leaking drains, or leaking water pipes. Unfortunately, storm drains and gullies to older property were rarely constructed well and soakaways will often have become blocked after the passage of time so that a point-discharge of water into the subsoil immediately adjoining the main walls will often occur. This can be damaging, especially to foundations on light sandy or silty soils which erode easily. Clients should be warned of the importance of maintaining both foulwater and stormwater gullies and drains in good order and to avoid any point-discharge of water into the subsoil adjoining the main walls.

Good building practice is to maintain foulwater drains in a watertight condition in normal use and to lay storm drains watertight for a distance of 3 m from any structure. After this distance, storm drains may be laid as open-jointed land drains if required, and may run to soakaways if alternative stormwater sewers are unavailable and the soil is suitably porous.

Clients will rarely give a thought to what happens to the storm and foulwater discharges from their property and their significance unless these points are described, and their importance to the structure is discussed. In practice, lack of access to storm drains will often preclude any description of the actual system by the surveyor.

Fig. 3.4 This settlement fracture has been poorly repaired and opened up again indicating continuing movement.

Assessment of damage

Foundation defects requiring a specific remedy are those which cause serious distortion or separation in the structure on a scale which would be unacceptable to a reasonable occupier. Specific remedies would

Fig. 3.5   A bulge fracture due to compression not related to foundations.

deal with foundations which are (a) not deep enough, or (b) not wide enough, or (c) not firm enough, for the particular subsoil conditions involved. The traditional remedy for such defects would be to underpin foundations in such a way as to make the overall structural support to the footings either deeper, wider or firmer as needed.

First one must consider those cases where such work would be unnecessary or premature. A number of specialist underpinning contractors exist who undertake traditional or system methods using interrupted foundations of various types. Such contractors will be commercially orientated and will wish to emphasise the advantages of underpinning using their own particular system. A surveyor will need to consider whether or not it is necessarily in the client's best interests to refer a problem to such a contractor rather than to a professionally independent structural engineer.

The first matter to be considered would be available evidence of actual past and present damage to the structure under review, having regard to age and all circumstances. Initially, any above-ground damage would need to be assessed, for example, cracks in walls (inside and outside), evidence of distortion to window and door openings, sloping floors and irregular ceilings. The Building Research Establishment (BRE) Current Paper 'Foundations for Low-rise Buildings', CP 61/78, and 'Assessment of Damage in Low-rise Buildings' in *BRE Digest* 251, make certain recommendations for the assessment of crack damage as follows:

Category 1   Degree: Very slight
Fine cracks up to 1 mm wide which can be treated during normal decoration.
Category 2   Degree: Slight
Cracks up to 5 mm wide, some external repointing may be required to ensure weather-tightness. Doors and windows may stick slightly.
Category 3   Degree: Moderate
Cracks between 5 and 15 mm wide, or a number of cracks up to 3 mm wide.

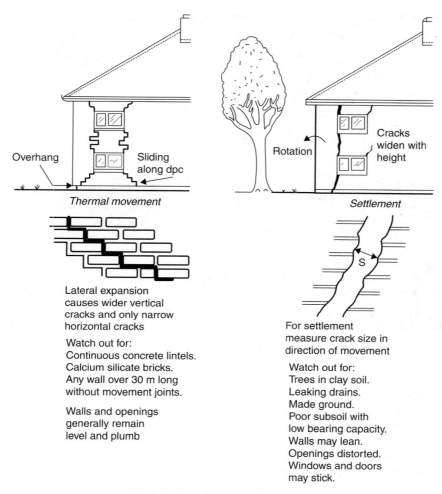

Overhang   Sliding along dpc

Thermal movement

Rotation   Cracks widen with height

Settlement

Lateral expansion causes wider vertical cracks and only narrow horizontal cracks

Watch out for:
Continuous concrete lintels.
Calcium silicate bricks.
Any wall over 30 m long without movement joints.

Walls and openings generally remain level and plumb

For settlement measure crack size in direction of movement

Watch out for:
Trees in clay soil.
Leaking drains.
Made ground.
Poor subsoil with low bearing capacity.
Walls may lean.
Openings distorted.
Windows and doors may stick.

Fig. 3.6 Assessing crack damage – 1.

Cracks require opening up and stitch bonding by a mason, with repointing to exterior brickwork. Doors and windows sticking. Service pipes may fracture. Weather-tightness often impaired.

Category 4   Degree: Severe

Cracks between 15 and 25 mm wide, but also depending upon numbers of cracks. Extensive repairs needed to sections of walls. Window and doorframes distorted, floors sloping, walls leaning or bulging. Some loss of bearing to beams. Service pipes disrupted.

Category 5   Degree: Very Severe

Usually 25 mm or more but depending on number of cracks. Requires major repair involving partial or complete rebuilding. Beams loose bearings. Walls lean badly and need shoring. Window glass cracks. Danger of instability.

Damage of Categories 1 and 2 proportions would not generally of itself justify any specific remedy, and a cosmetic repair should be sufficient. Categories 1 and 2 damage in circumstances where other adverse features of the structure or site exist may, however, be a foretaste of more serious problems to come. Most major structural fractures start with a hairline crack.

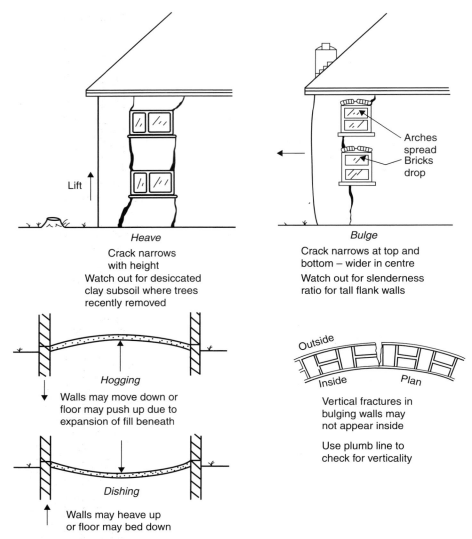

Fig. 3.7 Assessing crack damage – 2.

Category 3 damage could require a specific remedy depending upon the circumstances. Certainly such damage would require periodic monitoring in order to confirm whether or not the structural movement is continuous and worsening.

Categories 4 and 5 damage will probably require a specific remedy in addition to above-ground repairs. There is a likely need for some type of underpinning in these cases.

It should be mentioned at this stage that some settlement in certain structures is quite normal, being due either to the initial consolidation of the building after construction or to long-term consolidation over the life cycle. A BRE Current Paper, 'Settlement of a Brick Dwellinghouse on Heavy Clay, 1951–1973', CP 37/74, on this topic covered a 20-year history of settlement in the narrow strip foundations of a typical house in Hemel Hempstead, Hertfordshire. This settlement was found to continue fairly steadily for the whole period (without causing any cracking to the walls) at rates of between 0.12 and 0.39 mm/year, with

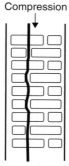

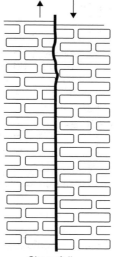

Overloading

Vertical cracks to
piers indicate
compression failure
with wall eventually
bursting laterally.

Watch out for:
Slender brick piers,
especially at
corners and bays
taking high loads.
Weak stonework.

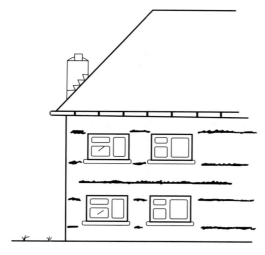

Rusting wall ties

Pattern of horizontal
cracking to mortar
joints in cavity wall.
There may be bulging
and separation of two
skins of brickwork.

Watch out for:
Any pre-war cavity walls.
Black ash mortar.
High exposure for driving rain.
1950s Finlock gutters with
leaking joints.

Cracks tend to appear
every six courses, this
being typical wall
tie spacing.

Shear failure

Often to bonding
between new and
pre-existing construction.

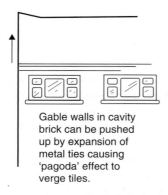

Gable walls in cavity
brick can be pushed
up by expansion of
metal ties causing
'pagoda' effect to
verge tiles.

Fig. 3.8 Assessing crack damage – 3.

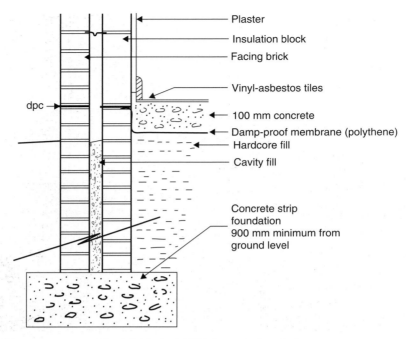

Fig. 3.9 Typical 1970 domestic foundation.

angular distortion of the foundations to 0.1%. More-brittle wall surfaces, especially surfaces with a smooth stucco, would have cracked as a result of such angular distortion causing Categories 1 or 2 damage, but of itself this would not have been significant.

The question of which category in which to place a specific observed area of damage is very much a matter of judgement. A large number of narrow cracks may, in some cases, be more significant than a few larger ones. The recommendations for categorising damage can be no more than a guide, and each property and area of damage has to be individually assessed on its merits. Having decided initially on the extent of the damage, advice must then be given.

There are three possible situations with which the surveyor may have to deal. First, acting for an intending purchaser and in this circumstance the client will have the choice of either buying the property or not buying it. If the client decides to buy he or she will probably have to do so within a reasonably short space of time, therefore there will be no opportunity for an extended period of observation. If there is any doubt about the significance of structural damage, it is clearly better *not* to buy the property and to look elsewhere. If then having had all the possibilities explained to him or her the client still wishes to buy, it is suggested that some adjustment of the purchase price might be recommended having regard to (a) the maximum likely cost of remedying the most severe defect likely to exist, and (b) the probability that such a remedy may be required having regard to the results of whatever investigations the client is prepared to undertake prior to buying.

The second situation occurs when the surveyor is consulted by an owner who is worried by signs of structural distress in his or her property. Advice given might then be (a) to do nothing, or (b) to subject the structure above ground to an extended period of observation using tell-tales and other indicators, or (c) in Categories 4 and 5 cases, to undertake temporary shoring and carry out trial excavations to determine the facts with a view to a subsequent specific remedy such as underpinning.

The third situation which can arise is where a building is inspected for a prospective mortgagee and signs of structural distress are found. In this situation the amount of the mortgage advance proposed, in

Fig. 3.10 Settlement or subsidence. The crack widens with height.

Fig. 3.11 Subsidence damage inside house.

Fig. 3.12 Subsidence damage outside house.

relation to the current market value, should be taken into account and a view expressed on the client's behalf of structural repair costs which could arise in relation to the probability of such repairs being needed and to the percentage advance involved. In many cases of lower percentage advances it would be reasonable to make such an advance on the security of a building suffering Categories 1, 2 or 3 damage. Categories 4 or 5 damage would require further investigation unless the mortgage advance was very low indeed. In addition to advising mortgagees in such circumstances the surveyor would be bound to bring such a situation to the attention of the borrower if he is to have a copy of the report, and the borrower should be advised to take further advice before proceeding, irrespective of the outcome of the mortgage application.

   A further point worthy of note in connection with structural damage due to subsidence, landslip or ground heave (normal insured risks in many building insurance policies) is that such a cover applies only to damage arising after the policy came into force. It follows that an existing building owner may be covered for such risks in respect of damage apparent in his building but that if he sells, the purchaser may *not* be covered if signs of damage were apparent before the sale. Once a purchaser has commissioned a surveyor's report which has described structural damage he may be placed in a different position to recover subsequent costs under his insurance policy. In such a situation it may be that the vendor could make a claim and agree the extent of remedial works required prior to any sale. A purchaser should not assume that, because he is insuring against such matters, minor cracks reported by his surveyor need not concern him since the very fact that the surveyor has reported the damage may have undermined the basis of any later claim.

Remedies

Damage caused by the action of tree roots and proved to be due solely to the presence of trees, and no other cause should not normally require underground repair. It should be sufficient to recommend that the offending trees be removed and that the site be given a number of wet winters to stabilise before the

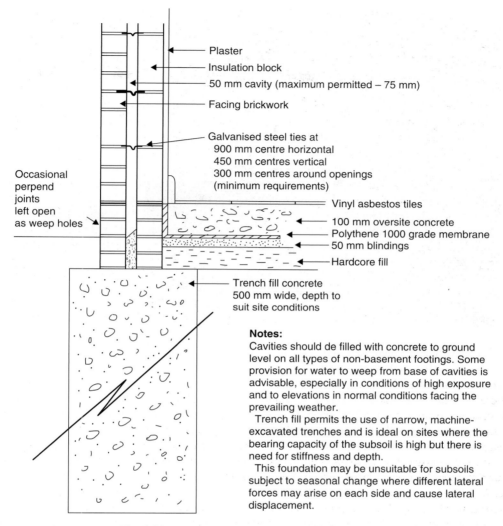

Plaster

Insulation block

50 mm cavity (maximum permitted – 75 mm)

Facing brickwork

Galvanised steel ties at
900 mm centre horizontal
450 mm centres vertical
300 mm centres around openings
(minimum requirements)

Occasional
perpend
joints
left open
as weep holes

Vinyl asbestos tiles

100 mm oversite concrete
Polythene 1000 grade membrane
50 mm blindings
Hardcore fill

Trench fill concrete
500 mm wide, depth to
suit site conditions

Notes:
Cavities should de filled with concrete to ground
level on all types of non-basement footings. Some
provision for water to weep from base of cavities is
advisable, especially in conditions of high exposure
and to elevations in normal conditions facing the
prevailing weather.
  Trench fill permits the use of narrow, machine-
excavated trenches and is ideal on sites where the
bearing capacity of the subsoil is high but there is
need for stiffness and depth.
  This foundation may be unsuitable for subsoils
subject to seasonal change where different lateral
forces may arise on each side and cause lateral
displacement.

Fig. 3.13 Modern deep strip or trench-fill foundation.

damage is made good by stitch-bonding cracked brickwork, repairing pointing and making good rendered surfaces in the normal way. If the offending trees cannot be removed, because they are either protected by tree preservation orders or in other ownership, then deeper foundation bases may be required
to accommodate present and future root activity. An interrupted foundation should be used to ensure that
expansion and shrinkage rates in the subsoil are the same on all sides of the foundation at different times
of the year. If trees are a contributory cause of damage but the foundations are also inadequate for other
reasons then an underground repair may be necessary in addition to the removal, pollarding or crowning
of offending trees.

  Traditional underpinning, which is an expensive and labour-intensive process, comprises the provision
of an additional foundation below the existing foundation. This may be achieved by alternately excavating
bases, usually 1 m^3 in extent, around the perimeter of the structure and under the internal load-bearing
walls. Each base is cast incorporating steel reinforcement married into adjoining alternate bases so that

a complete continuous foundation is provided under the existing foundation to whatever dimensions are specified. When the bases have hardened, the space between new and old foundations is filled by dry packing with a hard cement mortar. During the course of such underpinning, allowances have to be made for diverting and repairing any services encountered. After the work is finished, a suitable period is allowed for consolidation of the structure on the new foundation, and then any cosmetic repairs, masonry repairs and levelling up are completed.

As a general rule such underpinning will be applied equally to all bearing walls, otherwise a danger of structural movement between those parts of the building which have been underpinned, and those parts which have not, arises. This may not apply on a sloping site where the underpinning depth may step up with the site slope and remain in the same stratum of subsoil.

Special difficulty arises with semi-detached or terraced structures in multi-ownership since if the general rule mentioned is to apply, the whole structure or terrace would need underpinning. In such situations a degree of engineering judgement is required and it may be necessary to undertake partial underpinning only. If so, it is recommended that the depth of the underpinning bases be graduated down from the point where non-underpinned walls end to avoid a sudden marked change in overall foundation depth at any specific point. If party walls have to be underpinned, hopefully this will be by agreement since although a party wall can be physically underpinned from one side only, the bases themselves will be trespassing beneath and adjoining ownership and permission is then needed. There is also a risk that partial underpinning of, say, a semi-detached house and the party wall will result in any future stresses being transferred to the adjoining house. This could cause movement relative to the party wall with consequential damage. Before advising on such works legal advice should be taken as to the implications and means by which the adjoining owner might be warned about the proposed works. A party wall may be underpinned by Notice under the London Building Acts (Amendment) Act 1929, part of the special code for party structures which applies in London now extended to England and Wales in The Party Wall etc. Act.

As an alternative to the traditional underpinning described, a number of specialist contractors now employ special systems to provide greater depths using a system of pads and beams. Pads are cast at corners and within the structure at the main point-loads to the considerable depths if necessary. The pads are then linked by narrow reinforced ground beams cast directly under the original foundations. Such systems reduce the overall amount of excavation necessary and provide interrupted foundations which could be recommended on certain types of shrinkable clay, especially, if live tree roots are present in close proximity.

If a surveyor is asked to deal with underpinning he or she should be satisfied that it is necessary in the circumstances that the shoring and other temporary support works will be adequate, and that the procedure for casting a new foundation will not subject the structure or any adjoining structure, to additional distress.

Generally, with traditional underpinning, bases should be cast leaving at least two equivalent areas of support between so that bases 1, 4 and 7 would be cast, allowed to harden, dry-packed completely and the packing set. Then bases 2, 5 and 8 might be excavated similarly, followed by bases 3, 6 and 9. In soft ground even greater spacing of excavations may be required to avoid further structural damage. In all cases excavations should be open for the least possible time since some subsoils deteriorate when exposed. All excavations should be adequately planked and strutted at all stages of the work to avoid lateral displacement of the adjacent subsoil and the bottom of any underpinning excavation should be carefully checked to ensure that it is sound, level and dry before concrete placement begins.

If it is suspected that the structure under review has been underpinned in the past, the reasons for this should be ascertained and details obtained if possible. Nowadays, underpinning works require local authority consent under the Building Regulations and most recent works should appear in their records. Before 1976, such work was regarded as outside the scope of the Building Regulations or bye laws and normally went unreported to the local authority.

Drains and foundations

During the course of the survey it may be found that drains pass beneath foundations and floors or a shallow drain may pass through the walls below ground level and beneath the floors.

Certain precautions have to be taken when building over drains. Occasionally, a structure may have been built over drains originally and this will be found in many terraced Victorian and Edwardian houses with private drains running under the buildings from rear to front. Often, extensions to an existing building will have been erected over the drains at a later stage and this will commonly be found in small, single-storey domestic extensions.

Hopefully a manhole may be found on each side of the structure where the drain passes beneath, and it should be possible to test the drain for leakage and to inspect the pipework for distortion using a mirror. Such drainwork should be laid with a compressible layer above and immediately over the pipework and with suitable lintels used to carry strip foundations or footings over the pipes without imposing a point load. Modern unplasticised polyvinyl chloride (PVCu) underground drain systems are sufficiently flexible to cope with slight distortion without damage but older vitrified earthenware pipework is brittle and could fracture under compression caused by foundations.

One important point which should be considered in drainwork and shallow foundations is the relative depths of the original excavations where these would have been made side by side. It is not uncommon to find that the invert level of drainwork is deeper than the base of strip foundations. If the two excavations proceeded together at the time the building was erected, or if the drains were laid later and if the excavations have been in close proximity, then there would have been a danger of lateral displacement of the subsoil into the lower excavation following the work and during the period of back-fill consolidation.

Fig. 3.14 Subsidence has affected the lower part of this wall. As a temporary measure raking shores have been placed against the wall and hardwood wedges driven into the open bed joints to prevent further movement of the construction above.

This could have an adverse effect on the subsoil bearing the weight of the adjoining foundation, and the effects may last for some years causing settlements in the structure adjoining the drains until finally the subsoil will have fully consolidated to a point where further movement will not be significant.

It is therefore recommended that invert depths of all drainwork near walls of buildings be checked where possible and especially if that drainwork runs alongside walls. Deep invert levels adjoining structures built with shallow foundations may indicate that subsoil consolidation around the back-filling could be taking place with adverse effects on that structure.

Fig. 3.15 Settlement or subsidence again with the crack widening with height.

Mining and filled ground

Some surveyors practise in mining areas which are either active or worked out. In localities subject to mining subsidence it may be necessary to employ specialist advice before reporting on buildings, and it is recommended that the surveyor's library should include all available information in the form of plans and maps showing areas subject to mining where these are known. Coal-mining areas are mainly affected by this problem but there are many other types of mining and some areas have a mining history which goes back to the Middle Ages and beyond. In such situations the local knowledge of the surveyor may prove invaluable.

In areas having a chalk subsoil old chalk-mine workings may exist and the chalk itself may be subject to the sudden unexpected appearance of deep shallow holes either connected with workings or due to subterranean collapse caused by groundwater erosion. There is no effective way in which a surveyor carrying out a normal survey can forecast such events. The only protection available for the property owner against such eventualities would be to carry the normal subsidence and landslip insurance cover.

In some localities chalk pits, gravel pits, sand pits and worked-out quarries may be encountered where materials have been removed and the workings filled using either hardcore or compressible fill material. Buildings erected on bad ground where unsuitable fill material has been used, or where suitable material has not fully consolidated, are often found to be subject to movement. This type of structural movement can be highly damaging, occasionally requiring the demolition of the building and often necessitating a special underpinning scheme making piles (or deep pads) to take foundation loads below the level of the fill. Due to the pressures to build on existing urban sites or reclaimed industrial land, the number of serious defects attributed to building on made ground appear to have increased in recent years.

An ordnance survey map of the area sometimes indicates old mines and workings, especially on 1/1250 and 1/2500 scales. This can be a useful source of information; also the Geological Survey Map (GSM) may show where such workings have taken place. Old maps of the area sometimes show detail which no longer appears on modern maps and the writers have found the old ordnance survey county series of large scale maps to be useful, sometimes in showing conditions on a particular site before it was developed. Many of the older county series maps date from the last century. The old 1/2500 scale maps are often very accurate and informative.

The vast majority of foundations will perform satisfactorily throughout the life of a building notwithstanding the apparently inexhaustible range of foundation problems which could, in theory, arise.

A great deal of superficial damage to buildings, especially cracking, is minor and due to normal long-term consolidation of the structure or minor differential settlement. This must be regarded as normal, especially in older buildings. Some damage may be wrongly diagnosed as due to foundation movements when other and more prosaic causes may be responsible, especially in the case of widely based structures where thermal movements may be expected and in recently completed buildings where normal shrinkages, bedding down of partition walls and the like, will produce considerable cracking which may be quite unrelated to conditions below ground.

Serious damage is that which may be diagnosed as progressive and which will require a structural repair in order to arrest further movement. Most cracking is minor and not progressive. In the case of progressive movement there will normally be indications of the cause, and this will often be obvious to one who has a knowledge of local soil conditions and the presence or otherwise of mining or made ground in the locality.

In certain circumstances, compensation is payable for damage caused by mining subsidence if this can be shown to arise from the activities of the National Coal Board. Subsidence is also normally an insured peril in a standard comprehensive household insurance policy, but not in all cases and frequently not at all in policies on commercial buildings. Insurance cover will not normally include defects which existed or originated before the policy was taken out.

Landslips and coastal erosion

The surveyor with local knowledge is often at an advantage in those localities where there are fault lines, slip planes or coastal erosion. Some districts are notorious for the damage caused to buildings by continuing shifts in the landscape; for example, at Ventnor on the Isle of Wight a local fault line (or 'vent') is responsible for a long history of structural damage to buildings constructed on the boundary between two geological masses. Slip planes are found between layers of clay or other material on hilly or undulating sites and one mass of clay will be sliding over a mass below causing highly damaging shifts to foundations (see Morgan v. Perry in Chapter 19).

If a building is located close to a cliff or headland then clearly it is necessary to consider the possibility of coastal erosion and if there is any doubt regarding the condition of any coastal defences or the stability of the local geology then the client would be well advised to steer clear of such a property.

Geological Survey Maps

The importance of checking details of the subsoil by reference to the GSM when necessary was emphasised in *Cormack* v. *Washbourne [1996] EGCS 196*. A survey had been carried out on a house in Knotty Green, Buckinghamshire. The surveyor did not consult the GSM which showed Reading clay overlaying the chalk. Subsequent to the survey serious faults developed in the house which had to be demolished and rebuilt, the problem being associated with both subsidence and heave of the clay soil. The case went to the Court of Appeal in December 1996. The Court held that the surveyor's failure to consult the GSM was causative of the plaintiff's damage because it was the presence of the Reading clay beds beneath the house, marked on the map, which was the cause of the damage.

Fig. 3.16 Builders will often use a very shallow foundation for a porch thinking that because the weight is low they do not need to excavate as deeply as for the main house; this is not so and the porch in this case will need to be rebuilt on deeper footings.

4
Walls

Bottom: Some man or other must present Wall; and let him have some plaster, or some loam, or
some rough-cast about him to signify wall; and let him hold his fingers thus, and through that
cranny shall Pyramus and Thisby whisper.

Midsummers-Nights Dream Act III
William Shakespeare

'Mud Huts' by the sea

In the village of Clovelly, on the north Devon coast, the steep cobbled streets wind down to the harbour
between the cottages at such an angle that vehicular traffic is no use and the local villagers use wooden
sledges to carry things up and down.

In places the lime mortar rendering to the cottages gets damaged by the sledges and breaks away, especially around the base at exposed corners, and an observant tourist will see that the material behind the
rendering is rather unusual – not brick, stone or wood but a reddish-brown clay mixed with chips of wood
and grit. These houses are constructed of cob, and ancient building material not dissimilar to the clay and
earth used to build the mud huts of Africa, or the adobes of Mexico, and which was used from the earliest recorded times in England for cheap housing in rural areas.

Cob is rarely used nowadays. The latest examples of cob probably date from before the First World
War, and most existing cob construction is much earlier than this. Surveyors may occasionally be asked
to survey or value a cottage built in cob however, or advise on the restoration of a cob building. An
important point to remember is that the walls must not become too dry since the material becomes pow-
dery and looses strength if attempts are made to insert damp-proof courses. Some moisture within the
walls is necessary to maintain adhesion.

On the other hand cob walls will also deteriorate rapidly if they are subjected to excessive dampness
so discharges from overflows and defective guttering can be very damaging. In the West Country most
cob properties originally had thatched roofs with the deep eaves/overhang providing good protection for
the walls. Modern tile or slate roofs with much shallower eaves will not provide the same protection
against driving rain.

Traditional cob construction is built on top of a brick or stone footing (called a pinning) which extends
up about 300 mm above ground level and provides a perimeter base on which the outer walls rest. The
wet cob mixture is then laid on the pinning layers of about 600 mm at a time with further layers added as
soon as the material beneath is reasonably dry. Construction is very rapid and the walls are left for a couple
of days to harden before being rendered inside and out with a lime plaster and finished with a lime wash
as decoration.

For more rapid construction the cob can be laid inside a rough shuttering of boards to prevent the wet mixture spreading as it dries under the weight of additional material being added on top.

Owners of cob buildings are advised to allow for planned maintenance but not to undertake any significant alterations. If the structure is satisfactory it is best left alone. In particular any enlargement of openings or cutting new openings in walls is unwise since this may weaken what is, inherently, an already weak form of construction. It is not normally possible to underpin a cob structure because it cannot cope with the stress.

For the surveyor identification of cob is not always easy but the walls will generally be very thick, typically 450 mm or more with a distinctly wavy and unevenly rendered surface. The best clues are likely to be found in the roof space, under stairs or beneath floor boards where exposed cob construction may be seen. Cob can be eaten away by masonry bees and other insects, and if the material is soft around the base mice and rats will burrow into it. Hence the importance of maintaining the rendered surfaces in good condition, free from cracks, and carrying out regular checks to confirm that the construction is vermin proof.

In Buckinghamshire a particular local form of cob known as witchert is found in a belt extending from the Oxfordshire border north-east to Aylesbury and beyond. Witchert consists mainly of decayed Portland limestone and clay mixed with water and chopped straw and was widely used for small village houses in the 17th and 18th centuries with a few examples dating from as late as 1920.

With witchert a plinth of rubble stones, known as grumplings, is constructed in a shallow trench and the witchert is laid on top of this in layers known as berries, each berry being about 600-mm high, with the sides trimmed with a sharp spade. The walls are then rendered and lime washed. This type of construction is also commonly used for garden walls in Buckinghamshire without any rendering but with a simple tile coping on top and garden walls constructed in this way are surprisingly long-lasting.

An earth wall needs good boots and a good hat to keep it dry (but not too dry). It also needs to be able to breathe so traditional lime render and plaster should be used with lime wash as a decoration. Modern hard cement mortars and gypsum plasters are unsuitable, as are impervious masonry paints.

There are a number of specialists, mostly based in the West Country, able to supply blocks of cob for use in the repair of cob walls with helpful advice available on their web sites. Information regarding the maintenance of witchert houses is available from the Department of Planning, Property and Construction Services at Aylesbury Vale District Council.

A modern variation of cob construction, used extensively abroad but as yet very little in the UK, is rammed earth. With rammed earth shuttering is erected and the local subsoil is used to mix with Portland cement or similar and water, and then compacted either manually or using machinery to produce a very dense wall with an interesting and potentially attractive textured finish. We are likely to hear more about rammed earth construction in the future along with a growing interest in sustainable building generally. The obvious advantage is that the principal material for the walls is excavated from the site where the house is being built.

Brick and stone

There is a severe shortage of good bricklayers and stone masons now due in part to the cyclical nature of the building industry and its failure to train adequately when times are hard in order to have sufficient skilled men (and women) for the good times. At the same time there is renewed interest in ornamental and decorative brickwork and stonework with many volume developers now imitating Victorian and Edwardian details on their new estates – to a somewhat mixed critical reaction from architects and the apparent approval of potential purchasers.

The choice of brick is important if a durable job is to be achieved. Sadly there was a lack of knowledge among speculative builders in the past and they often used unsuitable bricks in exposed locations. As a general rule ordinary-quality bricks (Common Flettons and the like) are suitable only for construction above damp-proof course and below protective eaves or other covering in conditions of normal weather exposure.

Thus in conditions of high exposure, for raised parapets and copings or below damp-proof course, special quality bricks should be specified. During a survey particular care is required in checking brickwork used for garden boundary walls and retaining walls since these are not covered by Building Regulations (unless the retaining wall supports the house) and are not therefore subject to any inspections by Building Inspectors or District Surveyors.

Builders will often use whatever bricks come to hand for this type of external wall construction with little thought to the question of damp-proof courses, tankings, copings or movement joints. As a consequence boundary and retaining walls suffering from frost damage, cracks and spalling are commonly encountered during subsequent surveys. Spalling brickwork may also be found to brick-on-edge parapets are other detail where ordinary-quality bricks were used rather than special quality.

In my student days bricks were either commons, semi-engineering or engineering type and these expressions are still used now within the trade or on-site. However in architectural and engineering offices now brick classification is far more technical with a whole variety of types available to specifiers, graded in terms of appearance, hardness, strength and resistance to frost.

In the London area some new construction is built using second-hand brickwork, usually London Stocks reclaimed from demolition sites. The yellow stocks are very hard and durable. Victorian bricks are ideal for reclamation because the soft lime mortar is easy to knock off and they can be cleaned up and graded on the demolition site very quickly. At this point any soft bricks are rejected together with bricks subject to long-term tar staining from chimney flues. One hundred year old London Stocks will be good for another 100 years at least however when the time comes they will not be reclaimed again because if they are set in a modern Portland cement mortar it will be virtually impossible to clean this off without damaging the bricks, such is the strength and adhesion of cement.

In some parts of the country local stone is used for new construction, or reconstituted stone which is basically concrete blocks incorporating stone dust and chips in the mix to mimic the appearance of the real thing. And, of course, there will be lots of old stone houses about which will need to be surveyed from time-to-time. Some notes on a selection of local building stones appear in Chapter 15.

Cavity walls

The cavity wall was a quantum leap in terms of brick wall construction when it was introduced for residential building between the wars. The idea of having two skins of brickwork, separated by an air gap to improve weather resistance and insulation, became increasingly common in the 1930s and almost universal for post-war housing. The idea was not new however; the Romans used cavity walls and cavity construction of various types was used in England in some localities in Victorian times with cast iron ties, special shaped bricks and other means used to tie the two skins of brick together.

The early cavity walls from the 1930s onwards were often over-ventilated at top and bottom so that the standard of thermal insulation was sometimes little better than a half-brick thick wall making these houses very cold in winter. In the 1960s and 1970s cavities were generally closed off at the top under the eaves and air-vents were sleeved. Insulation block, rather than brick, became the usual material for the inner skin providing improved thermal insulation. And from the 1980s onwards we began filling cavities either partly or wholly with insulation much to the horror of many older surveyors who were taught the importance – above all else – of keeping the cavities clear.

Unfortunately the wall ties holding the two skins together are not always as durable as the bricks themselves and wall tie failure will be a particular problem with the early cavity walls especially to tall gable-ends facing the prevailing weather, walls built in black ash mortar (where industrial ash rather than sand is used as an aggregate) and where the thickness of galvanising to the wall ties is poor.

The life of a steel wall tie is dependent upon the zinc coating applied during galvanising and if the zinc is worn away by exposure to moisture over a long period, or chemical attack, the steel rusts. Rust can occupy up to seven times the volume of the steel which is replaced so as the ties rust they expand cracking the brickwork at the bed joints and allowing more weather penetration, a feature associated especially with the traditional once-twisted fishtail type of galvanised tie.

Mortars and bonding

The purposes of the mortar are to hold the bricks apart rather than stick them together and to seal the joints in order to provide weather protection. In traditional brick and stone construction the mortars should generally be weaker than the bricks or stones to which they are applied for reasons discussed in more detail in Chapter 21. In older buildings lime mortars will generally have been used whereas in most modern construction built since the 1930s Portland cement mortars, with or without the addition of some lime, will be usual and these modern mortars are often very strong and hard.

In my view the most attractive brick walls are faced in Flemish bond, although English bond is slightly stronger and there are a number of other bonds in less common use details of which are found in the usual construction text books. The outer skin of a cavity wall in simple stretcher bond is very bland and uninteresting in comparison. Cavity walls can be built to replicate Flemish bond using snapped headers to the outer skin but the bricklayers want extra money for this because it is much more time-consuming.

Lintels

No lintel is needed if a properly formed arch is used over an opening since the brickwork is then self-supporting. In all other cases the surveyor needs to check for signs of cracked or displaced wall construction over openings which may signify inadequate or failing lintels. In modern houses the ubiquitous galvanised steel lintels are used. In earlier post-war construction we find concrete lintels, either pre-cast or cast *in situ*, incorporating some steel reinforcement. In pre-war buildings timber lintels may be found and timber lintels set behind brickwork or stonework will be quite usual in most period buildings.

Where there is a ground floor bay window to a Victorian or Edwardian house the wall above will usually rest on a timber bressumer and these tend to sag with age and may give way completely as a consequence of rot or beetle attack. The first signs of bressumer failure will be some downward settlement of the brickwork over the bay and in these circumstances some opening-up of the bay roof construction would be advised. If the bressumer is in poor condition replacement with a rolled steel joist or universal steel beam may be necessary.

Some bay windows, especially to typical inter-war semi-detached houses, do not have lintels and the upper bay walls and roof gables rest on the window frames. No problems may be experienced until the owner decides to change the windows and is encouraged by a salesman to fit the latest unplasticised polyvinyl chloride (PVCu) double glazing. The new windows will have to incorporate tubular steel or similar supports at the outer corners but it is surprising how often this is not done and problems associated with replacement double glazing in bay windows are very common. During the survey the surveyor should always take a critical look at any replacement windows to see if they are square and consider whether the wall above is likely to be adequately supported.

Wall layout

During the survey the thickness, length and locations of walls around and within the building should be noted, including any abutting or connecting walls which are part of adjoining premises or boundaries. Those walls which are load bearing and non-load bearing can be distinguished.

In simple two-storey dwelling house construction, as in the ubiquitous inter-war semi there is usually a central spine wall which supports not only its own weight but also the loading from floor and ceiling joists above which span front-to-rear and probably also the strutting to the roof purlins. An opening may have been formed in such a wall to form a 'through lounge' and a section of steelwork will then be required resting on suitable bearings at each end to support the loading from above. To comply with Building Regulations such steelwork needs to be encased in plasterboard. The presence of a rolled-steel joist or universal steel beam can be confirmed easily enough using a metal detector but the bearings and actual steel dimensions cannot be checked without exposure.

With a roughly dimensioned sketch of the room layout on each floor one can easily see which walls rise continuously within the building and which rest eccentrically on the floors. In the case of the inter-war semi it is likely that some of the first floor walls rest on the floors supported by a double joist or similar if joists run parallel to the wall or on a sole plate over the boarding if the wall is at right angles to the joist run. In post-war construction greater use has been made of fully spanned roof trusses so that often all the partition walls at first floor level rest on the floors and do not correspond to the location of ground floor walls.

The internal partitions may be of blockwork imposing a considerable loading on the floor construction, or they may be lightweight timber stud partitions clad in plasterboard imposing fairly minimal loads. Sometimes considerable deflection of the floor construction will take place with the heavier type

Fig. 4.1 Raking shores used to support bay suffering from subsidence pending repair. Note the tree.

of blockwork partitions resting on timber floor construction resulting in distorted door openings on the landing and plaster cracks over door heads. Point loads of copper cylinders in airing cupboards often exacerbate this type of movement to internal partitions.

When surveying older houses from the Edwardian era and earlier the importance of taking full notes showing the arrangement of the walls and their construction is even greater since the internal walls to upper stories are invariably timber framed stud partitions clad in lath and plaster or similar having very little resistance to tension at the points of contact with the outer walls. A careful examination of the outer walls is then needed to see if there is evidence of bulging and leaning especially at tall gable-ends.

In old buildings the chimney breasts act as piers providing valuable stiffness and their removal is not recommended unless a careful check on the adjacent construction shows quite clearly that the slenderness ratios (SRs) of the remaining walls will be satisfactory.

Leaning or bulging walls may be restrained using tie bars with the S-plates or X-plates visible on the outer faces of walls. The tie bars themselves usually run under the floors and many of these were installed in London during and just after the Second World War to deal with walls rendered unstable by bomb damage. It is obvious that the effectiveness of a tie bar will be determined by the condition of the steel rod and the secure fixing at the other end – matters which can only be confirmed for certain by exposure.

Slenderness ratio

SR is a useful concept to use in assessing the structural stability of buildings under review. It is recommended that the standard to apply to old buildings is Regulation D15 and Schedule 7 of the old Building Regulations for structural works of bricks, blocks or plain concrete, and using Schedule 7 it is not necessary to calculate loads or wall thicknesses required, the rules of the schedule being sufficient as a check on the stability of the construction.

Schedule 7 may be applied to walls which form part of any storey of a building other than a basement, basement storey walls needing a structural check by calculation in accordance with BSI CP 111:1970 should this be considered necessary.

It is assumed for this purpose that point-loads are properly distributed and that walls are properly bonded and solidly put together. There is an overall height limit of 12 m for the use of these rules (above which CP 111 should be used) and also a limit of $3 \, \text{kN/m}^2$ as an imposed floor load. This is well above the imposed loads likely in residential construction but which could well be exceeded in other types of building, in which case CP 111 and special calculation would be required.

The table to Rule 7 sets out certain minimum thicknesses for external walls and separating walls, in each case assuming that walls will have at each end either a pier, buttress, buttressing wall or chimney unless the wall concerned forms part of a bay window and is less than 2.5 m in both height and length. This table is set out below:

Height of wall	Length of wall	Minimum thickness of wall
Not exceeding 3.6 m	Any length	200 mm throughout its height
Exceeding 3.6 m and not exceeding 9 m	Not exceeding 9 m	200 mm throughout its height
Exceeding 3.6 m and not exceeding 9 m	Exceeding 9 m	300 mm for the height of one storey and 200 mm for the remainder
Exceeding 9 m and not exceeding 12 m	Not exceeding 9 m	300 mm for the height of one storey and 200 mm for the remainder
Exceeding 9 m and not exceeding 12 m	Exceeding 9 m	300 mm for the height of two storeys and 200 mm for the remainder

For cavity walls the thickness of the wall should be at least 250 mm or the thickness required for a solid wall under Rule 7, plus the width of the cavity, except in the case of the upper storey of a two-storey dwelling house where the inner skin may be 75 mm (giving an overall width of 225 mm) subject to certain special rules including a doubling of the normal number of wall ties, roof loads being carried partly by the outer skin, a strong mortar being used, and the wall being not more than 8-m long and 3-m high (5 m high to a gable).

For the purpose of Rule 7, the heights are measured from the base of the wall immediately above the foundation or footing and not from ground or finished floor levels.

Walls are regarded for this purpose as being divided into separate lengths by piers, buttresses, chimneys and buttressing walls. These separate lengths are measured centre-to-centre of the piers, etc. provided that piers and buttressing walls themselves meet the necessary requirements and are fully bonded.

Figure 4.2 shows part of an end-of-terrace Victorian cottage. Note that the brick cross-wall construction is separated from the flank wall on both levels by door openings so that little or no effective cross-wall bonding exists. The stud cross-wall is of no benefit to the stiffness of the flank wall either. The flank wall is 225-mm thick and divided into three lengths, these being 3, 5 and 3 m, respectively, this flank wall being carried up to form a gable-end with an effective height exceeding 3.6 m.

If the chimney breasts are removed a single length of 11 m of wall exceeding 3.6-m high results which should be a minimum of 300-mm thick at ground floor level. Since this is not so, the removal of the chimney breasts is not recommended; they add essential stiffness to the flank wall construction. If soft lime mortars are used, the removal of buttressing from the flank wall may still be ill-advised, even if a check by reference to Rule 7 raises no technical objection, because the SR by reference to CP 111 is excessive.

SRs should be calculated in accordance with the formula laid down in CP 111. For two-storey residential buildings the Code recommends that the SR for walls built in cement mortars should not exceed 24 or 18 if built in the older type of lime mortars (12 if more than two storeys).

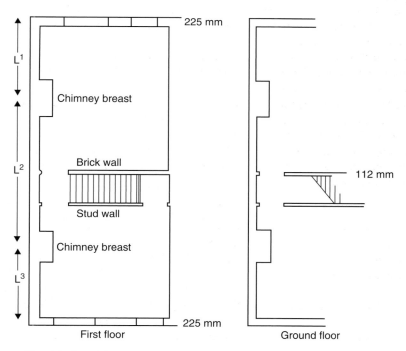

Fig. 4.2 Ground and first floor plans of end-of-terrace Victorian cottage.

The SR is determined by the length and height of the wall, the lateral restraint provided and the nature of the construction. A superficial examination of the overall construction will show if a building 'feels' right. Experience will tell whether a problem may arise having regard to the layout of the cross-wall construction and age of the structure. Newer practitioners may find it necessary to check such details by approximate calculation until they have acquired the necessary experience.

Having assessed the condition of the main walls and the manner in which they are likely to cope with the forces imposed upon them, the surveyor should then consider the other factors determining whether those walls are likely to be satisfactory or not. These are: (a) their resistance to dampness rising up, penetrating through or percolating down the walls, (b) their thermal efficiency and (c) in the case of separating walls (or party walls), their soundproofing properties. It is proposed to deal with each of these factors in turn.

Dampness

The building under review should be examined and compared with an equivalent modern building constructed to current standards. The extent that it falls short of modern standards should then be considered and the client advised accordingly. A good standard would be cavity-wall construction with clear, unobstructed cavities, having damp-proof courses at the base of walls at least 150 mm (two brick courses) above surrounding ground levels and with vertical and horizontal damp-proof courses at all points of contact between the outer and inner skin, and damp-proof cavity trays at all points of bridging. Additionally, there should be an impervious damp-proof membrane in solid floors contiguous with the damp-proof courses in all walls. By such means the passage of moisture into the interior from the exterior, or the site, is avoided.

The fact that a particular building is modern will not, of itself, ensure that a good standard is met since, in practice, poor detailing of cavity walls, obstructed cavities, punctured membranes and other faults abound. Nevertheless, a good yardstick to apply is that described. In so far as any construction under review fails to meet this standard, then this should be pointed out to the client, and many older buildings will not reach this standard.

Before about 1938, much residential construction was in solid brickwork or stonework and the main barrier against weather penetration in such construction is the condition and pointing of the external brick or stone face. In such circumstances this should be explained to the client, who should be advised to maintain the exterior surfaces in good order in future years and make a particular point of checking periodically for gutter and downpipe leaks. Leaking overflows will also cause considerable damage to internal plasterwork if neglected since the moisture can easily penetrate directly to the inside.

The clearance of 150 mm between exterior levels and damp-proof courses is recommended in order to prevent gutter-splash reaching the wall above damp-proof course and to allow some margin of clearance. Experience indicates that the ground levels around houses tend to rise by about 300 mm every 100 years, due to the addition of soil to gardens and the laying of new paving over old. It is essential to resist this trend.

When there is insufficient clearance to damp-proof courses it is recommended that ground levels be lowered. A general lowering of levels is best with the surplus soil or paving being carted away. Alternatively, a channel may be formed alongside the wall, at least 150 mm wide, suitably formed with cementwork sides and benching, and having a fall-away from the main walls and adequate drainage to suitable storm gullies. A channel which fails to drain adequately is worse than useless since it will actually exacerbate any dampness. The brickwork then exposed at the base of the wall should be made good as necessary by pointing, cutting in new bricks or rendering to a suitable standard.

Vertical damp-proof courses of slates set into a rendered plinth are a traditional remedy. They are inserted between the high exterior levels and the walls where damp-proof course clearances are inadequate or the courses themselves are bridged. Such vertical courses often prove ineffective in the long run as the slates become porous and the bedding mortar between slates is rarely impervious to moisture. Any vertical slate damp-proof courses found between high exterior ground levels and the external walls of older buildings should be regarded with a critical eye and consideration given to a general lowering of levels or tanking applied to any at basement or semi-basement.

Damp-proof courses become porous with age and rising damp results. Specialist contractors now provide remedial services either by chemical injection of the walls or electro-osmosis to prevent capillary attraction in the walls, or by inserting new damp-proof courses in sections using a bricksaw to cut out chases in the walls, generally of about 1 m at a time. Electro-osmosis is now generally considered to be less effective compared with other methods and one large specialist firm has discontinued its use. Electro-osmotic damp-proof courses should be given special consideration if encountered during a survey.

Before allowing a client to employ such a specialist contractor it should be confirmed that the dampness found actually is due to rising damp. Many cases of diagnosed rising damp are not due to failure of the damp-proof courses at all but from penetration or bridging. It is significant that guarantees offered by specialist contractors will normally exclude liability for dampness due to causes other than moisture rising up the walls by capillary action. Before employing a specialist contractor it is recommended that obvious defects likely to cause dampness should first be eliminated and the structure subjected to a period of observation, if this is possible. It will often be found that dampness can be cured by such simple steps alone.

Having said this, occasions will undoubtedly arise when dampness may be diagnosed as rising damp due to failure to damp-proof courses. In such circumstances all main walls will need attention since the damp-proof courses will be of the same age and type unless parts of the structure date from different periods. The provision of new damp-proof courses in only a part of a building and leaving the original damp-proof courses in other parts may be unwise since further work in the other areas will be necessary in due course if the damp-proof courses are failing due to their type and age.

Specialist contractors are commercial men seeking work and there is nothing wrong with that. As a professional, however, the surveyor should be able to say whether or not the specialist's proposals are unnecessary or premature.

Bridging of damp-proof courses can arise from inside the building as well as from outside. Frequently there is a build-up of rubble and debris under suspended timber floors. Whenever such floors have to be exposed for building works or services it is recommended that the over-site areas be cleared out and swept, and a check be made that all vents are free from blockage. Solid floors in older buildings may also bridge the damp-proof courses. The use of membranes in solid floors was not general practice in pre-1939 construction, and in halls, kitchens and sculleries quarry or ceramic tiles were often laid directly on the screed and thus provided a passage for moisture to rise up into the base of adjoining walls. It will often be found that plaster at the base of walls next to such floors has been affected by damp and a high moisture-meter reading obtained due to this damp transference.

In modern construction the failure to ensure that membranes are lapped over or under damp-proof courses and sealed at all joints can give rise to local bridging. Polythene can also tear or puncture, especially if laid under a concrete slab on gritty or stony blindings. Often, 500-grade polythene has been used as a cheaper substitute for the more durable 1200-grade which the writer would recommend for this purpose.

Thermoplastic tiles and woodblocks, commonly employed in houses of the 1950s, were laid in a bituminous backing which was itself the damp-proof membrane. Sometimes such tiles or blocks are removed by a do-it-yourself owner and replaced with other surfaces, the bituminous backing being disturbed or removed altogether. Modern vinyl tiles or welded sheet vinyl flooring cannot be laid with spot adhesive on a floor surface unless this surface rests over a suitable membrane. Failure to observe this will result in the floor screed sweating under flooring and causing it to lift and loosen. The bituminous backing to tiles and woodblocks should be contiguous with the damp-proof course in the walls but this may not always be so and can also be a source of bridging and moisture transference into the wall above damp-proof course level.

In older houses with damp solid floors or with timber floors in poor condition below recommended levels a useful remedy can be to have them replaced by new solid floors at a higher level. This can often

Fig. 4.3 The Victorians used special-shaped bricks as wall ties in their cavity walls.

be done usefully in kitchen and scullery areas where ceiling heights are sufficient. The ceiling height should not be reduced below 2.3 mm (the old Building Regulation minimum), preferably higher. The new floor can then incorporate a damp-proof membrane lapped at the sides and contiguous with the damp-proof courses in the walls. These should first be exposed by chasing the wall plaster and removing skirting boards as necessary. The client will then be able to lay vinyl tiles, sheet flooring or similar surfaces on the new screed in the knowledge that the floor will be dry and will not transfer moisture into the surrounding walls.

Apart from normal moisture-meter tests which should be made at selected points around the base of the main walls and at other points at risk, the wall surfaces should also be examined for signs of deterioration in the plaster, staining to wallpaper and mould growth, all signs of dampness (although not necessarily rising damp). Rust stains from skirting board nail heads are a good indicator of moisture behind the skirting board.

One final point regarding dampness in modern buildings is that in many post-1945 houses the pipework for plumbing and heating systems has been buried under solid floor screeds above the membranes and if an isolated point of dampness is found at the base of a wall, especially an internal wall, it is well worth while investigating the possibility of a leak in the system under the floor. If a serious leak is taking place from a heating system this can be confirmed by tying up the header tank ball valve, marking the side of the header tank at the existing level and then checking this level periodically to see if it drops. Any rapid loss of water from the header tank will indicate a leak to the central heating pipework in the floors which will have to be traced and remedied.

Thermal insulation

The thermal insulation requirements in current Building Regulations are reasonable standards to adopt and if the building under review is likely to be less satisfactory then the client should be warned.

Generally, a cavity wall with insulation block inner skin and cavity fill is a minimum present-day requirement. Older solid wall construction will allow greater heat losses, as will all-brick cavity construction of the 1930s and 1950s, especially if the cavities are vented and open at the top as was common practice during those times. The standard of insulation of an all-brick cavity wall with well-ventilated cavity is, in practice, not much better than that of a half-brick wall.

Still considering thermal insulation, it should also be borne in mind that some buildings are inherently less efficient than others. Bungalows have twice the roof area for the same floor space as two-storey houses and a bungalow with a number of projecting wings and bays will be presenting a large external surface area, especially if it is detached. The ideal structural shape from this point of view is a square box which encloses the maximum volume for the least external wall area. In view of the importance of thermal insulation, this topic should be covered by the surveyor in the report and if any steps should be taken to improve insulation, then these should be incorporated in the report.

All-brick cavity walls could be provided with sleeved vents through the cavities and the cavities sealed at the tops. Cavity-wall fillings may be employed. Solid walls can be improved by the provision of 3-mm polystyrene insulation under wallpaper on inside faces of exterior walls, and this will compensate to some extent for the lack of cavity construction. 'False' walls may be employed inside exterior walls with dry linings on battens and insulation, or plasterboard and dab.

Sound insulation

Current Building Regulations provide that party walls must be of 225-mm brickwork or dense blockwork, plastered both sides, to give the necessary sound-proofing. In practice, the older, denser party walls, often incorporating substantial chimney breasts, have been found to be more satisfactory in this respect than the more modern party walls. In a Building Research Establishment current paper, 'Field Measurement of the Sound Insulation of Plastered Solid Brick Walls' (BRE CP 37/78, March 1978), it was found that in houses built during the 1970s the test results indicated that walls constructed to meet the deemed-to-satisfy specification of the Building Regulations actually failed to meet that standard in practice in about 35% of cases.

The reasons for this include the fact that voids may arise in brickwork and blockwork if the workmanship is poor and frogs not fully filled. Chases for electrical and aerial conduits, etc. often corresponding in location on both sides of a party wall, may allow local channels for sound to pass. In so far as it is possible to do so, the surveyor should comment upon the degree of sound insulation in separating walls, although in many cases this will not be possible.

Wall design

Figure 4.4 shows a sketch section through an unrendered solid brick wall of a type commonly encountered in Victorian and Edwardian construction having a solid internal floor covered with quarry tiles and no effective damp-proof membrane. A risk of damp penetration through the wall and dampness rising by capillary action via bridging arises with this type of construction. Wet- or dry-rot to grounds and skirting boards in such a location is likely if dampness is present. Simple remedies such as dealing with defective rainwater equipment and lowering ground levels may suffice but it will not be possible to confirm whether the slate damp-proof courses are effective or not since they cannot function properly anyway until the obvious problems are dealt with. Floors may be re-surfaced with a new covering of two good coats of 'Synthaprufe' or similar bituminous coating, taking care to cut back old tiles, to expose the damp-proof courses and to finish the bituminous backing against the slates to ensure a continuous damp-proof course and membrane. Pointing and soft brickwork should also be attended to as required.

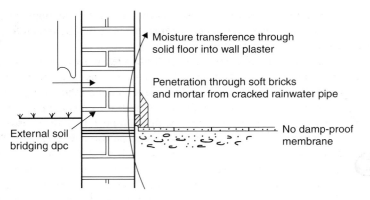

Moisture transference through
solid floor into wall plaster

Penetration through soft bricks
and mortar from cracked rainwater pipe

External soil
bridging dpc

No damp-proof
membrane

Fig. 4.4 Bridging of slate damp-proof course in unrendered, solid brick wall.

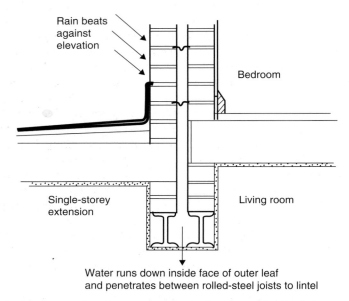

Rain beats
against
elevation

Bedroom

Single-storey
extension

Living room

Water runs down inside face of outer leaf
and penetrates between rolled-steel joists to lintel

Fig. 4.5 Lack of damp-proof cavity tray to cavity wall over opening for living-room extension.

The standard of thermal insulation of such a wall will be poor and heat loss is excessive by modern standards.

Figure 4.5 shows a sketch section through a cavity wall where a living room has been extended and two small rolled-steel joists inserted. In such a location, especially in conditions of high exposure or normal exposure facing the prevailing weather (usually south west), a damp-proof cavity tray should be provided, lapped over the exterior roof flashing and a few open perpends or weepholes allowed for so that any water running down the inside face of the outer skin is picked up and taken to the outside over the roof line. Similar provision is required in a cavity wall where a subsidiary pitched roof joins the outer skin using a conventional flashing and where the cavity continues downwards to terminate at lintels over rooms below.

The provision of rolled-steel joists and other lintel work in residential extensions of this type is often badly done but the detail can rarely be checked. Apart from the provision of proper damp-proof cavity trays and weeping perpend joints as necessary, it is essential that proper bearings for joists are provided and that for all but the very short spans, the rolled-steel sections should be bolted together with spacers to provide improved stiffness and resistance to torsion.

Occasionally, walls will be found which are rendered both sides and which will appear from measurement to be of brickwork. Care should be taken in making assumptions since such walls may be of either solid or cavity concrete construction, formed originally in sections using shuttering. There was a vogue for such construction for a time in certain localities in the inter-War period, and more recently the development of such construction using no-fines concrete has been popular. The nature of the construction of a rendered wall may often be confirmed by checking inside air-vents and at the top of walls around the perimeter of the roof space. Good building practice requires that wall cavities be closed at the top but in practice open cavities are often found, especially in older cavity walls. It is possible to look down such cavities from within the roof space in some cases in order to check for blockages, bridged wall ties and other defects as well as to confirm the nature of the construction used for the outer and inner skins. Inspection within open wall cavities or other concealed areas can be undertaken using endoscopes or borescopes which require only narrow access through an air-vent grill or a small drilled hole in one skin – however endoscopes cannot be used to any advantage where the cavity is filled with insulation. This equipment is not generally used as a routine test in all surveys but is particularly suitable for checking older cavity wall construction for corroded wall ties. Requiring a drilled hole about 10 mm in diameter, the endoscope can be used to inspect floor and roof voids, ducts and casings, is illuminated and can be linked to a 35-mm camera.

As regards cavity-wall construction, whether of brickwork, stonework, blockwork or a combination of materials, the overall stiffness of the construction relies upon the two skins being held together by effective wall ties. Since some cavity walls of a considerable age may now be encountered, many being more than 60 years old, it is worth considering how effective the wall ties are now likely to be.

Current Building Regulations require wall ties specified or manufactured to British Standard, BS 1243:1978, 'Metal Ties for Cavity-wall Construction' or an equivalent including non-ferrous or stainless-steel wall ties. Galvanised wall ties usually take the form of either vertical twist, double triangle or butterfly types, and the durability of the ties is dependent upon the zinc coating given during galvanising. The possibility of corrosion and the possible resulting instability of cavity walls is a known hazard and some reports of structural failure of cavity walls for this reason have been made, especially in cases where black-ash mortar has been used. In a British Research Establishment (BRE) information paper, 'The performance of cavity-wall ties' (BRE Information Paper IP 4/81, April 1981), it was concluded that vertical-twist ties may lose their zinc coating within 23 to 46 years and wire ties in similar conditions within 13 to 26 years, assuming zinc coating to the current British Standards. The use of wire ties was considered unsatisfactory, and more durable ties were recommended although the majority of existing buildings were regarded as sufficiently robust for widespread problems to be unlikely.

Faced with cavity-wall construction and black-ash mortar (where an ash-fine aggregate has been used instead of sand) the surveyor should beware of corroded ties. And faced with any cavity-wall construction over 50 years old should be on guard against the possibility that wall ties may have corroded to the point where there is no effective connection between the two wall skins. Conventional low-rise cavity walls will normally remain quite stable, even in the absence of effective ties, due to stiffness afforded by bonded cross-wall construction inside and abutments over, under and around openings, these abutments having only limited resistance to tension between skins but good resistance against compression. Corrosion of ties could be more significant in low-rise construction with little bracing and in high-rise buildings which have masonry infill panels with galvanised metal ties.

The possibility should be borne in mind that galvanising of connecting ties used for certain types of pre-fabricated construction may be poor. Surveyors will occasionally have to report on local authority houses (some now privately owned having been purchased by the tenants) constructed by using various forms of pre-cast concrete slabs and pre-fabricated panels. A number of different systems were used during the

Fig. 4.6 Inadequate cover for steel reinforcement in concrete lintel. Rust can take up to seven times the space of the steel it replaces causing bursting to the concrete surface.

Fig. 4.7 Drill holes at the base of the wall indicate that a chemical injection damp-proof course has been added.

post-war municipal housing programmes. Variations in such construction will also be encountered in commercial and industrial structures including pre-cast concrete panels fixed using bolted connections and spacers with galvanising to bolts and fittings often of a poor standard. In addition to problems associated with rusting of galvanised fixings, some systems incorporated a framework of tubular supports which are prone to corrosion and the pre-cast concrete panels themselves may be subject to the rusting of the reinforcement bars inside, this causing disintegration of the panels.

Prompted by reports of deterioration to prefabricated reinforced concrete houses the BRE has published reports on individual PRC house types including Boot, Cornish, Unit, Orlit, Unity, Wates, Wollaway, Reema, Parkinson Framed, Stent, Dorran, Myton, Newland, Tarran, Underdown, Winget, Ayrshire County Council (Lindsay) and Whitson–Fairhurst construction. Findings are summarised in Information Papers, IP 16/83 and IP 10/84 available from BRE. As part of an on-going programme the BRE intend to report on all the main types in time.

The main problem is corrosion of reinforcement which expands causing concrete to crack and spall and investigations revealed that the reinforced concrete components are gradually deteriorating. While the great majority of houses studied were found to be in structurally sound condition there was a wide range in the rates of deterioration both between and within types, and there are no practicable techniques which can halt deterioration altogether. It was advised that houses of this type be regularly inspected and adequately maintained to prolong their remaining life by measures such as are described in BRE Digest 265, 'The repair of reinforced concrete'.

Monitoring walls

Occasionally walls will be either quite new or have surfaces recently covered in such a way that evidence of past damage will not be discernible. In these cases procedures will be available for monitoring the performance of the walls and sometimes, in an initial report, it may be necessary to recommend such procedures.

One of these recommended procedures is the accurate monitoring of cracked damage in distressed buildings, this generally being undertaken using calibrated tell-tales, these being rigid indicators fixed over crack walls and showing measured horizontal and vertical movements over a period of time. Tell-tales should be given reference numbers and all measurements of movement logged at regular intervals in a permanent logbook. After a suitable period has elapsed, the information collected will be useful in showing the extent and direction of structural movement and whether or not it is seasonally variable.

Movements in walls due to thermal expansion and contraction can show a distinct pattern of measurement, with conditions in hot and cold weather differing. Movement in walls due to progressive settlement of parts of a structure relative to other parts will show a general progressive overall measurement. Structural settlements followed by heave due to shrinkage and swelling of clay soils will produce distinct seasonal patterns. Expansion due to chemical deterioration of concrete members may be found to be more noticeable during periods of wet weather. The variety of movements which may be caused by subsidences and landslips of the subsoil are considerable, and by accurate monitoring over a period of time it is possible to confirm overall downward movements of parts of a structure, progressive leaning and tilting of walls, rotational movements of projecting wings and other structural manifestations.

Interpretation of crack monitoring results should be undertaken with care since it is sometimes easy to jump to wrong conclusions. Movements on shrinkable clays are particularly tricky to interpret since what appears to be a downward settlement of part of a building could in fact be stability with the other part moving up due to clay swell. The use of an external datum may be needed to confirm the point.

Clearly not all clients will be in a position to benefit from the monitoring of crack damage since if the client is a prospective purchaser he or she may need to decide quickly whether or not to purchase a building. If the client is proposing to take a repairing lease he or she may have no time to spare for lengthy structural monitoring. Similarly, prospective mortgagees will be unlikely themselves to be able to

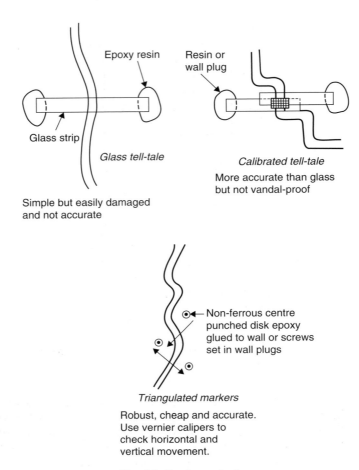

Epoxy resin

Resin or
wall plug

Glass strip

Glass tell-tale

Calibrated tell-tale

More accurate than glass
but not vandal-proof

Simple but easily damaged
and not accurate

Non-ferrous centre
punched disk epoxy
glued to wall or screws
set in wall plugs

Triangulated markers

Robust, cheap and accurate.
Use vernier calipers to
check horizontal and
vertical movement.

Fig. 4.8 Crack monitoring.

commission structural monitorings of walls prior to confirming any mortgage advance although they may be able to require the prospective mortgagor to undertake this and allow a suitable period for him or her so to do prior to confirming their offer. This situation is unlikely to be helpful to a prospective mortgagor who himself is an intending purchaser and under pressure to exchange binding contracts.

One feature of crack damage occasionally encountered in Victorian and Edwardian construction of a cheaper type is that some solid walls were built without crossbonding and snapped headers used to the outer face, such walls having the appearance of 225-mm solid brickwork but, in fact, comprising two skins of 112-mm brick with a continuous lime mortar joint rising vertically in the centre. Curious crack patterns, and leaning and bulging of such walls may be found showing different patterns to inner and outer faces. This construction was used to minimise cost by saving all facing bricks for outside and in practice may be found to develop instability with age, especially where tall gable-ends are concerned.

Retaining and freestanding walls

In addition to the main walls of buildings, surveyors often have to assess exterior walls such as retaining walls, freestanding walls or parapet walls (generally of brickwork or stonework) which form adjuncts to or extensions of the main structure. Unfortunately such walls are frequently badly built due to lack of

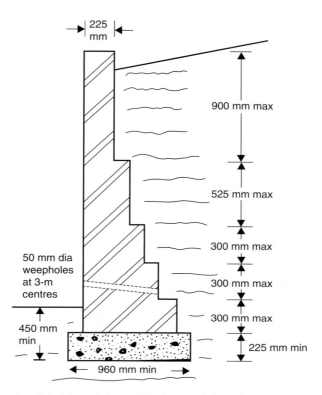

Fig. 4.9 Typical design for all-brick retaining wall, using special-quality brickwork which does not require tanking or weathering. The wall is stepped back in half-brick increments. Backward-sloping drain-holes prevent staining down the face of the wall from very small quantities of groundwater discharging over a long period. Note that ordinary-quality brickwork requires special tanking, damp-proof courses and weathering, otherwise frost damage will result.

understanding of the principles involved and (in most cases) the fact that they are outside Building Control Legislation.

Such walls are examined to see whether or not they are plumb and level. The overall construction should be considered in relation to the importance of the walls and the requirements of their construction. Clearly, retaining walls may be very important indeed to a structure on a sloping site and may be essential to the stability of that site and thus of the structure.

The first design consideration is whether or not the walls under review are allowed adequate provision for movement. Generally, movement joints in brick retaining and freestanding walls are desirable at about 10-m intervals to cope with normal thermal expansion and contraction, and occasional movements which occur as a result of chemical conversion in concretes and mortars. Movement joints at greater frequency may be recommended for freestanding walls on sites where the subsoil may be unstable and where, for reasons of economy, a shallow foundation is necessary. On shrinkable clays, for example, garden walls and the like could be provided with movement joints at 3- to 5-m centres, obviating later cracking in the brickwork due to seasonal ground movements.

The second design consideration is whether in the case of retaining walls the construction itself is adequate. Where a retaining wall is essential to the stability of a structure a consulting engineer's report should be obtained if the construction standard is doubtful. This will generally involve some site investigation

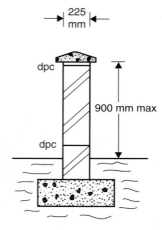

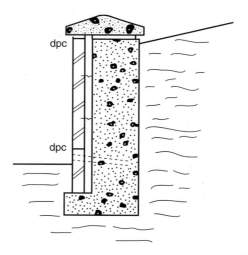

Fig. 4.10 Freestanding wall construction. Throated weathering projecting at least 45 mm, plus damp-proof courses at top and bottom, allow ordinary-quality brickwork to be used for normal exposure conditions, but special-quality bricks are necessary for footings below damp-proof course, (dpc).

Fig. 4.11 Mass concrete retaining wall with brick face.

behind the wall. In Figure 4.11 a cross-section of a brick retaining wall is shown. It will be noted that 225-mm brickwork should be used only for retained heights up to 900 mm, and that progressively thicker wall construction is required as the height of the wall increases. Regard should be given to the bearing pressure values of the retained ground, which may require special consideration. Figure 4.10 shows a typical freestanding wall. It should be noted that 225-mm brickwork is a minimum thickness and that piers or buttresses or, alternatively, thicker overall construction is required for heights greater than 900 mm. A check on the SR is recommended in doubtful cases, paying particular regard to the strength of the mortar, which is often poor, exterior brickwork and stonework, especially if this is old.

The third design consideration is whether the wall material will deteriorate should groundwater be able to enter the wall, and taking into account conditions of exposure. Special-quality brickwork or stonework resistant to frost damage and chemical attack from groundwater will be normally be suitable.

Fig. 4.12 BISF (British Iron and Steel Federation) Houses.

Fig. 4.13 The paving is too high. It should be lowered to at least 150 mm (2 brick courses) below the damp-proof course or internal floor level whichever is the lower, with a slope away from the wall.

Ordinary-quality brickwork or stonework which may not resist such deterioration should not be used in conditions of high exposure. In the case of normal exposure, ordinary-quality brickwork or stonework may be used provided that certain precautions are taken, these being the provision of tanking to retaining walls, and horizontal damp-proof courses at the tops and bottoms of walls, also the provision of adequate copings with throated overhangs. In Figure 4.11 a sketch of a mass concrete retaining wall with a facing of ordinary-quality brickwork is shown. Below damp-proof course at the base a special-quality brick footing is necessary. In Figures 4.10 and 4.11 the damp-proof course is shown as a line. For most situations however a simple bituminous felt or plastics damp-proof course should not be used due to the weak bonding line which results and the risk of overturning failure of the wall – a better solution is to use two courses of engineering brick to provide a damp-proof course with uninterrupted mortar joints within the construction.

Bulging walls

A bulge may arise for a variety of reasons including rotation of the foundation. A very small tilt to the foundation can cause the wall to lean out whilst remaining tied in at a higher level resulting in a bulge. Movements of this type in old walls built in soft lime mortar will often be accommodated without causing cracks.

Fig. 4.14 Hand made red facing bricks set in lime mortar to English Bond with a two-course slate damp proof course. Good ground clearance below the damp proof course.

5
Floors

We thought that the ramps at the doorways were for someone with a wheelchair. It was only after moving in that we realised that all the floors have sunk by three or four inches which is why a second set of skirting boards has been added on top of the original skirting boards.

Extract from letter from potential client to surveyor

Some years ago I was asked to advise on a claim being made against a surveyor in respect of his failure to note in his report that the ground floor slabs had sunk. The house had been built in the 1950s on a sloping site to a fairly cheap specification and as a consequence of the failure to consolidate the fill material under the concrete floors, and the use of unsuitable fill material, the floors had settled significantly and unevenly throughout the ground floor rooms.

Prior to sale the vendor had levelled up the floors so that under the fitted carpets and vinyl flooring throughout the ground floor there was a surface of chipboard, finished dead level at the base of the skirtings. Under the chipboard various layers of plywood and other material had been used to level up the floors, and beneath this was the original surface of woodblocks and thermoplastic tiles.

The surveyor concerned, undertaking a Building Survey, had rolled back the carpet in a number of places and noted the chipboard surface. He had also checked the floors with a spirit level. And he had concluded that the floors were fine. What he had not done was to question why a modern surface of chipboard had been added. The purchaser decided that he wanted to restore the original wood block as an appearance finish and then discovered that all the floors would need to be relaid and some pressure grouting of the fill material would be needed to stabilise the floors.

So if you see something odd, which is obviously not part of the original construction, it is always a good idea to ask why it was done and what problem was it intended to remedy.

Identifying the nature and condition of floors often presents a particular problem during a Building Survey inspection since the property may be occupied and the floors may be covered with fitted carpet, vinyl flooring or laminated wood flooring. A vendor will typically be reluctant to permit valuable floor coverings to be disturbed. A surveyor may then be guided by the feel of the floor underfoot, the degree of bounce to timber floors and factors such as poor sub-floor ventilation or surrounding dampness which may indicate that all is not well. Depths of floor joists or slabs can be checked at changes in levels or within stairwells.

Suspended timber floors

Suspended timber ground floors may be of the traditional type supported on intermediate sleeper walls or may be of the fully-spanned type. With either type it is clearly essential that the decking and the supporting timbers are of adequate type and size for the spans involved and a check by reference to the standards laid down in Building Regulations is recommended, using the appropriate approved document. The old 1976 Regulations contained a helpful table in Schedule 6 and this is reproduced below since it provides a useful standard to apply to older residential buildings.

For surveyors still familiar with imperial measurements, a rough rule of thumb for use in old buildings is that the depth of the joist in inches should not be less than the half the span in feet plus one. So for a span of 12 ft: 7″ × 2″ joists at 16″ centres would be a minimum for residential loading. For a span of 10 ft: 6″ × 2″. For a span of 14 ft: 8″ × 2″. For a span of 16 ft: 9″ × 3″ (joist width should be increased so as not to be less than a quarter of the depth).

In older houses the surveyor may encounter uneven and sloping floors, some with considerable gradients when floor joist are over-spanned, either being so originally or having had intermediate support removed from below. An overall gradient in one direction will, of course, not be due to deflection but will indicate

Schedule 6, Table 1, 1976 Building Regulations (applicable from 1976 to 1985 – now replaced by Approved Documents under current Regulations).

Floor joists for dead loads of not more than 25 kg/m^2 excluding the mass of the joist

Size of schedule joist (mm)	Maximum span to centre of bearings (m)		
	400 mm spacing	450 mm spacing	600 mm spacing
38 × 75	1.03	0.93	0.71
38 × 100	1.74	1.57	1.21
38 × 125	2.50	2.31	1.81
38 × 150	2.99	2.83	2.46
38 × 175	3.48	3.29	2.86
38 × 200	3.96	3.75	3.26
38 × 225	4.44	4.20	3.66
44 × 75	1.18	1.06	0.81
44 × 100	1.97	1.78	1.38
44 × 125	2.62	2.52	2.05
44 × 150	3.13	3.02	2.64
44 × 175	3.65	3.51	3.07
44 × 200	4.16	4.50	3.93
50 × 75	1.33	1.19	0.92
50 × 100	2.10	1.99	1.55
50 × 125	2.73	2.63	2.29
50 × 150	3.26	3.14	2.81
50 × 175	3.80	3.66	3.27
50 × 200	4.33	4.17	3.72
50 × 225	4.85	4.68	4.18
63 × 150	3.51	3.38	3.09
63 × 175	4.08	3.93	3.59
63 × 200	4.65	4.48	4.10
63 × 225	5.21	5.03	4.60
75 × 200	4.90	4.93	4.33
75 × 225	5.49	5.30	4.85

some more fundamental movement in the bearings to the floor joists. Faced with a floor which slopes in one direction only, the surveyor should confirm whether or not this is due to failure of the joist bearings or to a downward settlement of the wall on which they bear. A wall plate, for example, which is eaten away by beetle attack or softened by rot may simply crush under the weight of the joist ends resting on it and a downward movement equal to the depth of the wall plate could occur. This is a particular danger in old structures with wall plates built into the damp wall construction.

Upper floors in traditional dwelling-house construction are normally suspended timber floors, fully spanned between ground floor walls and supporting certain point loads such as airing cupboards and tanks, blockwork or stud partitions and the like. A check should be made on such floors similar to that recommended for ground floors, especially having regard to the possibility of over-spanned joists in older buildings.

For bearings greater than 100 mm a notional bearing of 100 mm may be used so that in most cases the maximum clear span will be 100 mm less than the span to bearing centres shown above.

The tables to the Building Regulations assume new timbers in good condition, classified according to the current British Standard (BS), or graded in accordance with certain minimum standards for a particular species. It will be appreciated that timbers in existing buildings, and especially older buildings, may not meet the grades of Group 11 specification laid down and some allowance for this may need to be made in applying Table 1 as a rule of thumb to check the adequacy or otherwise of floor joists, particularly where the span is marginal.

In domestic construction the most commonly found joists will be 50 mm × 150 mm, 50 mm × 175 mm and 50 mm × 200 mm, generally at 400-mm centres, with maximum permissible clear spans of 3.16, 3.7 and 4.23 m, respectively (i.e. the table spans less the distance to bearing centres on notional 100-mm bearings).

In domestic construction which has suspended timber ground floors resting on intermediate sleeper walls, the joists will normally be found to be 100 mm × 50 mm and the requirements of the old Building Regulations would assume clear spans between sleeper walls of 2 m, assuming 400-mm spacing and dead loads not exceeding 25 kg/m². It follows that ground floor construction with spans between sleeper walls exceeding 2 m clear may be over-spanned and this is a useful avenue of investigation for the surveyor to explore if the floors appear to flex unduly underfoot. In some newer properties where joist sizes have been calculated to comply with the Building Regulations rather than use of tables, such unacceptable flexing may often occur and can prove embarrassing, particularly, for example, in dining rooms where walking across the floor can make cutlery rattle.

Intermediate strutting to suspended timber floors is required where the span exceeds 2.5 m, using herringbone or other strutting to prevent undue warping or torsional distortion of the joists. The *Registered House–Builders' Handbook* recommends and requires one set of such strutting for spans of between 2.5 and 4.5 m and additional strutting for greater spans. It is important that the strutting be neatly done and tightly fixed otherwise it will not prove effective. In the event of strutting being absent in spans over 2.5 m, the surveyor may have to consider whether or not any undue warping or distortion of the joists has resulted, and recommendations to this effect are required. In practice, intermediate strutting is often found to have been omitted with no apparent ill effects.

It is important that bearings and floor joists be so constructed as to eliminate the possibility of moisture transference from adjoining masonry and from oversite into timbers. In older buildings, joists are often built into the outer walls where the construction is in solid brickwork and a likelihood of rot attack to joist ends then arises, especially if the wall concerned faces the prevailing weather or is in a porous condition (see Figure 21.1).

In some older buildings the joists will rest on wall plates which are themselves built into the outer walls or fixed to the face of walls, and the possibility of decay and even complete disintegration of the wall plate arises, causing serious loss of bearing to the joists.

It may also be found that sleeper walls or other means of support lack effective damp-proof courses under sleeper plates and joists, thus giving rise to decay and insect attack to the timbers. Commonly the air-space under a suspended timber ground floor will be found to contain large quantities of debris left over from the original construction or later building works, and this can cause moisture transference into timbers from the oversite and also block ventilation.

The provision of adequate underfloor ventilation to a suspended timber floor which is above an oversite in contact with ground moisture is essential since lack of good ventilation can give rise to condensation on the timbers and resultant decay. A careful check must be made for any 'dead' areas of air-space under such floors which may not be adequately ventilated. Generally a good standard of underfloor ventilation to a suspended timber ground floor in domestic construction would be 225 mm × 75 mm terra cotta airbricks, sleeved through wall cavities, sited at 1.5-mm centres. Iron grilles are not recommended since eventually they rust and permit the access of vermin. A vent should be provided at or near each extremity of the floor void to prevent areas of 'dead' air.

Solid floors

Solid slab ground floors of concrete on hardcore are more durable in the long term than suspended timber ground floors. In older buildings where timber floors are in poor condition, replacement with a solid concrete floor is often to be recommended for a number of reasons, especially if the adjoining ground levels are high and it is difficult to provide the necessary clearances to air-vents.

The main problems likely to arise with solid floors at ground-floor level will be due either to settlement of the floor slab or rising dampness. Settlement and dishing of post-war solid concrete floors due to inadequate consolidation of the hardcore filling beneath or the use of unsuitable fill material is a common problem, so common in fact that the National House-Building Council (NHBC) list it as being the major source of complaint in houses covered by their scheme.

On reasonably level sites, the NHBC will still approve solid concrete ground floors on a hardcore base and, in theory at least, such construction should be quite suitable for most ground conditions provided that the standard of workmanship is satisfactory. Most problems arise when workmanship is poor and site supervision lacking.

On sloping sites where a deep fill would be required under concrete ground floors, the NHBC now require suspended-floor construction to be used where fill would exceed 600 mm in depth. On a sloping site it may be that a mixture of solid concrete on fill and suspended timber (or concrete) floors would be necessary to comply with this requirement (NHBC Practice Note 6, 'Suspended-floor construction for dwellings on deep-fill sites').

Unless the laying of fill material is supervised constantly and thoroughly it is virtually impossible to ensure that consolidation is adequate and that all material used is suitable.

Occasionally problems may arise due to the use of fill material which is still chemically active, such as cases where industrial waste material or ash blinding is used. It is essential that all fill material be chemically inert, otherwise damage to floor slabs from expansion or contraction of the fill are possible, and chemical attack to the concrete and surrounding masonry could occur.

It is also possible for a solid concrete ground floor slab to be built over ground affected by existing leaking drains or for such drain leaks to arise after construction. Occasionally old land drains may exist and continue to discharge groundwater into the subsoil under the slab, thus causing, erosion.

When inspecting a building with a solid concrete ground floor the general floor levels and amount of deflection which may be present in relation to the age of the floor should be considered. Some initial consolidation is often inevitable. If the floor has been in place for some years and the extent of any deflection is acceptable (say up to 0.003 of the span), this may be regarded as a matter for mere comment. If the deflection is greater then it may be necessary to consider undertaking investigations under the slab to confirm the nature and quality of the fill material. A core sample or small sample inspection hole is the best means of achieving this result. The sample material should be examined and, if necessary, a laboratory report obtained.

When taking samples care must be taken to check for excessive moisture under the floor. Under normal conditions both the fill material and any subsoil under a floor should be nearly dry. Excessive amounts of water in samples may indicate a plumbing or drain leak under the floor and this could be one cause of the problem.

Where solid concrete floor slabs have settled due to inadequate consolidation of the original fill, a remedy may be required to stiffen the fill. This can be carried out, without taking up the floor, by using pressure–grouting techniques. Floor surfaces can then be re-screeded level, taking care to ensure the integrity of damp-proof membranes and also checking that membranes remain contiguous with the damp-proof course in the surrounding walls in order to prevent moisture transference.

An examination around the perimeter of a floor is always to be recommended since this could reveal any gaps which may have opened up below skirting boards. If the surveyor is satisfied that the skirting boards are at their original height, a measurement of the gaps will indicate the extent of any downward movement around that perimeter. If quarter-round beading or similar additional trim has been applied at the base of skirtings, this may indicate that gaps have appeared and subsequently been covered. This feature should place the surveyor on his or her guard over the possibility of settlement in that floor. New skirting boards may have been applied and this could be due to a number of causes, including the possibility of such settlement in the floor surface.

Floors should be checked for level using a spirit level where a visual inspection suggests a slope. Some surveyors even carry a marble and use this to see if solid floor surfaces are uneven. However, since most floors are uneven to a greater or lesser degree, the fact that a marble may roll on the floor will not of itself be a means of accurately gauging the extent of such unevenness. A long spirit level is much better. In a larger building it may be necessary to take levels using a dumpy level or water level in order to assess the degree of unevenness in the floors.

A series of spot levels will often reveal considerable fluctuations in a wide-based structure which are not apparent visually.

In good construction, timber and suspended concrete ground floors will be separated physically from any contact with the site or structure below damp-proof course line, and floor surfaces should therefore be dry. Solid concrete ground floors are made dry by the provision of a suitable damp-proof membrane, and in the absence of an effective membrane the surfaces of such floors become damp due to moisture rising up from the site, especially if the surfaces are covered with carpets or sheet flooring.

Damp-proof membranes may be located below the floor slab, in which case polythene film of at least 1000 gauge, and preferably 1200 gauge, should have been used, or alternatively, a bitumen sheet to BS 743. It is essential that all joints in such membranes be lapped and sealed, and that the membranes are carried up around the perimeter and lapped under the damp-proof courses in the surrounding walls so as to provide a completely imperforate barrier against rising moisture. Lapping should be at least 300 mm. Such membranes should have been laid over a suitable blinding of inert ash or sand, free from any sharp protrusions which could puncture the membrane.

Damp-proof membranes in sandwich construction may be located within the slab and could comprise polythene film or bitumen sheet (as before described), hot-applied mastic asphalt, pitch or bitumen, 0.6 mm minimum thickness of cold-applied bituminous solutions, or composite polythene and bitumen self-adhesive. In such construction the concrete slab is first allowed to dry thoroughly, and the membrane applied to the surface and lapped up to the damp-proof courses as necessary. A surface screed is then applied which, to avoid 'shelling' or lamination, should be a minimum of 50 mm depth.

Membranes also may be located on the floor surface in the form of hot-applied mastic asphalt, pitch mastic or cold-applied pitch/epoxy resins. In older construction hot-applied mastic will generally be found as a base for woodblocks or thermoplastic tiles, the use of polythene and other modern materials not becoming general practice until post-1960.

In pre-1939 and older constructions, solid floors were often not provided with any damp-proof membranes at all and the surveyor will often encounter flagstoned floors, quarry-tile surfaces or ceramic tile surfaces of various types and ages which may be performing adequately when fully exposed but which would immediately begin to 'sweat' if covered with tiles, sheet flooring or carpeting. In advising a client in such circumstances, and before modern surface materials may be applied, it is essential to explain the significance of the modern practice of providing membranes and the special attention required to older floors which lack such membranes.

For the first half of this century most housewives were content with a kitchen equipped solely with a sink and draining board, gas-point for an oven and a quarry-tile floor which could be periodically washed down with soap and water and polished on Sundays. Quarry tiles are not impervious to moisture but they present a reasonably adequate barrier and the need for a dry floor surface was not important in those days. Nowadays a wife would wish to see a modern kitchen installed in an older house in order to make it acceptable and the old quarry tiles replaced by vinyl tiles or sheet vinyl floorings. It is essential that any new floor surfaces be laid over a suitable membrane which is continuous with the damp-proof courses in the surrounding walls. To achieve this it may be necessary to remove skirting boards together with wall and floor material at the skirtings in order to expose damp-proof course in the walls and to seal the new membranes against these damp-proof course to avoid moisture transference.

Flagstone floors may be lifted and re-laid over a suitable membrane if flagstones are required as a finish, again paying particular attention to the need to joint the membrane to the surrounding damp-proof course.

In commercial structures, solid floors may be laid as dry floors or without any membranes. Again, provided that the floor surfaces are exposed, it will often be found that concrete floor will remain reasonably dry on the surface even if no membrane is provided, but once the surfaces are covered they will sweat. The needs of the client may require consideration since certain types of storage and warehousing, for example, would require dry conditions and dry floor surfaces.

Concrete floors in commercial structures can be dusty and a suitable treatment of concrete paints or special sealants to reduce surface dusting may be recommended. For many types of manufacture a dust-free atmosphere is essential especially where precision work or electronics are concerned.

Where a garage is either an integral part of a house or is attached to it with a communicating doorway into the main accommodation, the garage floor will normally be a solid concrete slab. This slab should be on a lower level than the main ground floors inside the house with a step up at any doorway of at least 150 mm. The purpose of this is to prevent burning petrol in a garage from spreading under the door and into the main house. Garage floors should also have a slight slope down towards the main entrance to allow water to run away.

Particular attention should be given for providing adequate drainage to paved areas outside garages, especially where driveways slope down towards the building, in order to prevent driving rain or stormwater running in through the main door opening. Generally a driveway which slopes down towards a garage should always be provided with a suitable stormwater gully and a suitable drainage fall to this gully. This is often found, in practice, not to be the case with the result that ponding of stormwater in driveways and garages occurs. In such cases a channel and grating across the threshold of the door and drained into a gully should be provided.

A floor may have a surface of boarding but may nevertheless be a concrete floor with the boards laid on timber battens on the concrete. This is commonly found in the blocks of Edwardian mansion flats that are a feature of our towns and cities, especially London. These boarded floors generally feel very hard underfoot with no bounce. The space beneath is typically about 50 mm – ideal for running heating pipes and other services.

Suspended concrete floors

Suspended concrete floor construction can take one of two possible forms, either pre-cast construction with or without a composite screed, or cast-*in-situ* construction. The latter is more commonly encountered in older buildings.

Cast-*in-situ* construction may be identified where the underside is exposed, mainly in commercial buildings, as the pattern left by shuttering is visible where it used to support the floor during curing.

When the floor construction is covered on all surfaces it will be impossible to confirm the nature of the construction without some special exposure or core samples being taken.

Pre-cast concrete floors appear in a wide variety of forms and slab sizes. It is more likely that defects due to conversion of concretes may arise in pre-cast construction, particularly if high-alumina cement (HAC) is present. HAC had an advantage in the production of pre-cast concrete members as it allowed a rapid drying time and quicker production. During the post-war period until 1974 a great deal of pre-cast floor slabwork and beams were made using HAC.

In February 1974 two roof beams in a school in Stepney collapsed and a major question arose as to the condition of beams and slabs in post-war buildings where HAC had been used. The Building Research Establishment (BRE) undertook urgent investigations to establish the degree of risk likely to arise in buildings with pre-cast concrete beams made with HAC. The results of their investigations were published in April 1975 in Current Paper, CP 34/75, 'High-alumina Cement Concrete in Buildings'.

The outcome, according to the BRE, was that the risk of structural failure to floors of up to a 5-m span made with pre-stressed pre-cast beams using HAC is very small.

Spans of more than 5 m are unlikely to be encountered in residential construction, including blocks of flats, except in exceptional cases, bearing in mind that cross-wall construction is generally used and the spans will normally be the shorter of any two room dimensions. In most cases a surveyor will be able to see when the existence of spans greater than 5 m is unlikely in floor slabs, although it is rather more common in blocks of flats where beams are used. In commercial buildings spans exceeding 5 m will often be encountered in both floor slabs and beams. If the building was erected between 1925 (when the use of HAC in the UK began) and 1975 (when it ceased) the possibility of chemical conversion of the concrete resulting in a loss of bearing capacity may then arise.

In cases where there is a risk of the presence of HAC in spans exceeding 5 m, the surveyor should recommend that a consulting engineer be asked to carry out tests and advise independently on the position.

Floors generally

Floors should therefore be reasonably level, capable of taking dead loads and superimposed loads without excessive deformation and should be dry. Floors may also have to accommodate service ducting in commercial buildings and for such purposes a floor with accessible ducting to accommodate pipework and cables is desirable. In dwelling houses it is preferable that central heating pipework and other services be laid so that easy access is possible – in solid floors ducts in screeds (above the damp-proof membrane) are preferred and in timber floors screwed traps at important access points are useful. It is also a good idea to incorporate dovetailed perimeter battens in screeded concrete floors (above the damp-proof membrane) to accommodate carpet fixings for fitted carpets.

In the absence of exposure it is unwise for a surveyor to make any positive statement regarding the condition of a floor. Defects are common especially in older timber floors and often give rise to complaints following the exposure of previously hidden construction.

Bear in mind also the poor sound proofing to timber floors in converted flats – a common cause of complaint from purchasers; the modern trend to have bare floor boards or laminated wood flooring can result in considerable noise nuisance for the occupant of the flat below. When you buy a flat with a simple joist and boarded floor above you rely for your quiet enjoyment of the property on the cooperation and consideration of your neighbour.

6

Roofs and chimneys

I looked into the loft myself and thought that it was rather light in there – I assumed that there was a skylight. I now realise that it was because so many of the slates were missing.

Note from client following receipt of report

The three most popular areas of activity for itinerant confidence tricksters (in the London area at least) appear to be tree lopping, driveway tarmacadam and roofs. Roof work is very popular amongst the con men. The victims tend to be the elderly and those least likely to want to go up a ladder to see for themselves.

In a typical case there will be a knock on the door, the householder will be advised that the man on the doorstep is a roofer working in the area who has seen serious problems with the chimney, flashings or whatever. Before the occupier has time to think men will be on the roof apparently doing something and a short time later there will be another knock on the door demanding payment, typically in cash. Elderly victims will be marched down to the bank to draw out their savings.

Despite all the press publicity and warnings from the police it still goes on. It goes without saying that surveyors will always advise their clients never to instruct a builder who just turns up on the doorstep or who has a van with no identification on it apart, perhaps, from a mobile phone number. Householders should always ask for a written quotation on proper-headed paper and if they are unsure whether the work needs doing or the quote is reasonable, they should take advice.

There are two basic types of roof. Pitched roofs with an inclination of 10° or more and flat roofs with an inclination of less than 10°. The results of roof failures can be serious since leakage of water into a building may give rise not only to the need for repair of the defect but also consequential losses. Special attention should be paid to the requirements of clients interested in non-residential buildings such as factory and warehouse space. If materials are to be stored in circumstances where roof defects could cause damage, particular attention should be given to this point.

Life of roof coverings

Most roof coverings have a reasonable and well-defined effective lifespan beyond which re-covering may be necessary or trouble may be anticipated. In practice, there is considerable variation in the quality of materials from different sources, especially traditional materials, but the following is a list of the principal roof-covering materials likely to be encountered and a guide to their lifespan.

	Years
Hot-bonded, built-up mineral felt flat roofs to BS 747, well constructed	10–15
Hot-bonded, built-up felt roofs constructed to lesser standards	3–10
Good quality flat lead roofs, well detailed and laid with rolls	80–100
Good quality flat copper roofs, well detailed	30–75
Zinc roofs, well laid with rolls, in areas not subject to chemical pollution	20–40
Zinc roofs in situations of chemical pollution	15–20
Asphalt, well laid and detailed, in two coats to BS CP 144, Part 4: 1970	20–40
Hand-made clay tiles with battens and sarking felt	60–70
Machine-cut clay tiles with battens and sarking felt	40–60
Plain concrete tiles, double lap with battens and sarking felt	50–90
Interlocking concrete tiles with battens and sarking felt	50–80
Interlocking Dutch or Belgian clay tiles	40–60
Asbestos-cement tiles or roofing sheets	20–25
Good quality Norfolk reed thatch	50–60
Straw thatch	20–25
Local stone slabs (depending upon porosity)	10–100
Good quality Welsh slates with battens and sarking felt	80–100
Poorer slates, subject to lamination with age	40–60
Shingles (depending upon quality and frequency of treatment)	5–30
Perspex, clear or obscure unplasticised polyvinyl chloride (PVCu) sheets, etc. (depending upon brittleness, etc.)	5–25

The surveyor should confirm during the survey, the nature of the roof coverings in various parts and assess their likely age and should then consider what advice may be given in each particular circumstance. Where roof coverings are likely to give problems in the foreseeable future the position should be explained to the client having regard to the possibility of either repair or renewal of these coverings. There will be many situations where re-covering might be recommended, but where the client could, if he or she so wished, extend the life of the roof more or less indefinitely by undertaking periodic overhauls and renewing individual defective slates or tiles as the necessity arises. Many clients are quite happy with an old roof covering which needs such periodic attention and will happily spend the money needed to renew it on something else. It seems quite reasonable in the circumstances for the surveyor merely to present the facts and leave it to the client to make the decision. Many examples of old slate and tile roofs will be encountered, still giving service well beyond the time scales indicated above.

Pitched roofs

Undersides of roof slopes will generally be lined with sarking felt in buildings erected post-1938 and in some cases earlier. Only a very limited view may be available of the underside of the roof covering if felt is loose or torn, and there will be no view of it at all if it is tightly and correctly laid. Unfelted roofs will normally allow an examination to be made of the underside of the covering, but where feather-edged boarding is used then very little access to the underside of the covering will exist. This is encountered mainly in 1920–1940 property.

It is helpful to examine debris which may have accumulated on ceilings in older roofs. Extensive deposits of clay-tile laminations and broken-tile nibs will indicate that the tiles are laminating and becoming porous. Slate laminations, sections of cracked slates and rusted nail heads will all indicate that a slate roof is approaching the end of its recommended lifespan, or that 'nail sickness' is causing otherwise sound slates to slip. Often a roof covering will have been replaced and the debris in the roof void will indicate the type of previous covering, a point which may be significant if a heavy covering has replaced a lighter one since roof timbers designed for the lighter covering could deflect under the weight of the heavier.

Pitched roofs without underfelt should be regarded nowadays as below the acceptable standard. Clients should be warned that the degree of weather-proofing, especially in driving rain and snow, will not be as good as with an underfelted roof and also that the roof void will become dusty and items stored there will therefore need some protection.

The scantling of the roof timbers should be checked. In older buildings, rafters may lack effective purlin support, or purlins may lack sufficient strutting. Sagging rafters or purlins should generally not be jacked back into line but recommendations to stiffen up the purlin or strut construction to a reasonable standard should be given. A cross-check by reference to Schedule 6 of the Building Regulations should be made, bearing in mind that the older timbers may not be of an equivalent standard, and it is better to over-size purlins, struts and the like in marginal cases.

Fig. 6.1 Cotswold stone roof with skylight and lead apron. Careful attention to detail is needed if leakage around skylights is to be avoided.

In older roofs a check for roof spread is advisable. The rafters should meet the ridge board tightly and if they have pulled away, this could indicate insufficient provision of collars, inadequate fixing of joists at the rafter feet, or insufficient binders to roof slopes which are at right angles to the main rafters.

In modern trussed-rafter and roof-truss systems certain important points are often overlooked. First, diagonal restraint is required to trussed roofs with gable-ends to prevent a parallelogrammatic, or 'racking', sideways movement of the trusses. Diagonal planking from the eaves at each corner to the ridge at the centre with two good galvanised nail fixings to each truss is essential, especially if the gable-ends themselves are of poor construction. Secondly, the end two or three trusses should be strapped with galvanised-steel strapping built into the gable-ends to prevent separation. Thirdly, no cutting or alteration of such truss construction should be undertaken, and if this is found to have been done, a warning should be given that the overall stability of the roof construction could have been seriously affected. Fourthly, only light storage on roof truss ties is advisable. The size and spacing of trussed rafters are not designed to accommodate heavy point loads.

Flat roofs

Most modern flat roof construction used in post-war dwellings is of hot-bonded, built-up mineral felt on timber decking, supported by timber joists and with a ceiling beneath. Built-up felt construction should normally comply with British Standard (BS) 747:1977 which sets out certain minimum standards for the materials to be used and the manner of laying. From a superficial inspection it is generally not possible to confirm whether or not the standards have been met. However, there are certain important matters which can be confirmed and borne in mind for future maintenance of these typical cold deck roofs.

First, for roofs situated over habitable rooms the felting should be in three layers laid in hot bitumen with the surface covered in bitumen-bedded chippings for protection from the sun. If chippings are not used the surface felt will lift and crack after only a short time, particularly on roofs with a south aspect. Periodic inspections should be carried out to confirm that the protective layer of chippings is in place and has not been washed away.

Secondly, the void underneath the decking must be provided with some ventilation. The decking itself may be of tongued-and-grooved or butt-jointed boards; it may comprise Stramit board, made of compressed straw, which was popular in the nineteen-fifties and nineteen-sixties; it may be of chipboard or some other type of manufactured boarding, either pre-felted, with only two additional layers of felt rather than three, or unfelted. The purpose of ventilation is to prevent an accumulation of trapped vapour under the decking which can otherwise cause lifting or 'blowing' of the felt. This may then indicate trapped water vapour in the void due to lack of ventilation. Ventilation may be provided quite easily by allowing for spacers under the perimeter fascia boards of small roofs and by setting vents into the surface at intervals where larger areas of flat roof are involved. Adequate ventilation is essential to prevent interstitial condensation.

Thirdly, the roof void should be protected from the entry of warm, damp air rising from rooms beneath (especially bathrooms and kitchens) by the use of suitable vapour barriers or vapour checks. This means that the ceilings beneath should be virtually sealed and airtight. If narrow ceiling cracks, holes or joints exist, particularly where electric wiring passes through, then warm, damp air will enter the void and condense on the underside of the cold decking in the roof void, a feature encountered mainly in wintertime. Polythene sheets may be used as vapour barriers provided that they are carefully lapped and sealed at all joints and are placed on the warm side of any insulation – generally this will mean directly over the ceiling plasterboard. Electric wiring should be run not through the sheets but under them so as to maintain the integrity of the vapour check. Foil-backed plasterboard is often used but in this case joints should be sealed. Metal foil, however, makes the best vapour barrier.

Fourthly, the roof should be insulated. Experience indicates that flat roofs tend to suffer greater heat losses in winter and heat gains in summer. This may be reduced by the provision of ceiling insulation which should be laid over the vapour barrier or vapour check but with an adequate ventilation gap to permit free circulation of air directly under the decking.

Fifthly, special care is needed to maintain the flat roof detail in good order since most problems start with small leaks at flashings, upstands and parapets and at points where openings are formed for soilpipes, flues and the like. Warm boiler flues will present a special problem since the built-up felt around them will be softened by the heat and the weather-sealing will often be deficient.

Where the roof forms a dished tray with an internal rainwater pipe, frequent inspections would be advisable to clear leaves and debris, especially if trees are near, since if the internal rainwater pipe blocks, the tray will fill with water which could then penetrate the decking at any weak points around the perimeter.

Well-constructed built-up felt roofs have a useful lifespan of 10 to 15 years, after which renewal of the covering will generally become essential. Occasionally a felt roof may last longer, especially if sheltered from the sun, but poorly-constructed felt roofs often have a shorter life.

A problem arises in giving advice on the action to be taken where a flat felt roof is weathertight but is of an age when re-covering may be anticipated. To some extent local repairs in the form of patching defective areas and renewing flashings and upstands are often possible and may prolong the overall life of the roof. The main problem arises from the fact that small leaks, often undetected for some time, will cause rapid deterioration of the decking and timbers beneath and destroy the effectiveness of insulation and the vapour barrier. In these circumstances the condition of decking and timbers cannot be accurately assessed without full exposure.

Stramit board decking will absorb moisture and convey it over a wide area by capillary action. If ventilation is poor, this warm, damp atmosphere will encourage wet- or dry-rot fungus to grow. Chipboard deckings will also absorb moisture and deteriorate rapidly, often without any moisture reaching the ceilings beneath to give warning to the occupants that all is not well above. Conventional butt-jointed or tongued-and-grooved boarding or exterior grade plywood sheeting are more durable. Any decay occasioned by small leaks in the surface covering has tended to be more localised than would be the case with chipboard or Stramit, although boarding is often more prone to outbreaks of wet- or dry-rot if it is unseasoned and ventilation is poor.

The re-covering of a flat felt roof in new built-up felt to BS 747 will be a fairly straightforward building operation and, for a small area, can be completed rapidly. If a small domestic roof is involved it is generally possible to strip off the old covering, have one or two layers of felt laid and the joints sealed in one day, thus avoiding the need for weather protection in the form of temporary canopies and scaffolding. However, if the roof area is large, or the decking and joists need repairs, then the roof will be exposed for some time and the cost will have to be increased to cover not only the need for weather protection but also the repairs to the joists and decking. In such circumstances, defective flat roofs can cause considerable inconvenience to a property owner and result in substantial expenditure to put things right.

The purpose, therefore, of periodic inspections of the roof surfaces and ancillary items is to pre-empt any trouble in the form of small leaks which could result in major expenditure later on.

Thatch

Occasionally a surveyor may be asked to report on a thatched roof. There are some thatched properties in suburban areas and a considerable number in rural locations, probably more than one would expect. Thatch gives considerable advantages as a roof covering. It has good insulation properties against both heat loss and noise, and the noise of rain on the surface goes unnoticed. Generally a deep eaves overhang is provided, obviating the need for gutter or rainwater pipes and rainwater is allowed to run off the eaves overhang well out from the main walls of the building.

Vermin infestation in the thatch can be a problem. Birds cause considerable damage during the nesting season, and squirrels and mice are very destructive to thatch once they have gained an entry. Many thatched roofs are covered with a fine galvanised steel wire mesh to prevent birds or vermin removing or burrowing into the straw or reeds but this can make access very difficult in the case of a thatch fire and a PVC or nylon mesh is now often used, melting when heat is applied.

Clients buying a house with a thatched roof should be warned that insurance premiums may be higher in respect of fire cover on both building and contents. A specific point should be made of notifying the insurers when a roof is thatched, this being a matter relevant to the risk. Certain obvious precautions should be taken by the property owner. He or she should not light garden fires near the building or use blow lamps for stripping paintwork near the roof and, in an urban location, some check should be made on Guy Fawkes' night and similar occasions and a hosepipe kept handy in case of ignition by fireworks or bonfire sparks.

Further, warning should be given concerning the estimated lifespan of the particular material. In the case of reed thatch about 60 years is typical but straw thatch has a much shorter life. During the life of the thatch the covering should be periodically inspected by a thatcher and may need an overhaul, generally taking the form of combing the surfaces, making good defective areas and renewing hips, ridges and details where these have deteriorated. A property owner should cultivate an association with a local firm of thatchers or, if no such firm is available, contact the nearest available rural locality where thatching skills are still practised.

When thatch reaches the end of its useful life it will need replacing in its entirety. Replacement may be undertaken with a different covering such as slates or tiles or with new thatch if thatchers are available. In many cases re-thatching will be realistic and not more expensive than tiling, bearing in mind that tiling will certainly involve new battens and sarking felt, renewal of or addition to rafters and the provision of gutters, rainwater pipes, fascias, soffits and different detailing around stacks. If the property is in a conservation area, or is a listed building needing special consent from the local planning authority, it may be that a change in the type of roof covering will not be permitted.

The fire risk associated with thatch will obviously be greater in older timber-framed buildings where the ceilings and walls may be of lath and plaster or similar construction which is itself a potential fire hazard in some circumstances. The need to maintain the electrical equipment in a property with a thatched roof in good order, and to test the wiring regularly, is self-evident. This is extremely important since when the thatch is renewed or repaired, the old thatch is often left lying in the roof void. This dry, brittle, dusty material is highly combustible, particularly in the case of straw thatch.

One final warning concerning thatched roofs is the prevalence of insect infestation – in particular furniture beetle and the death watch beetle where oak timbers are involved. There is hardly a thatched roof in the West Country which does not contain some form of insect infestation, usually *Anobium punctatum*.

Roof coverings in general

Lead is a traditional covering for both flat and pitched roofs with the life dependent upon the lead thickness adopted. Heavy cast lead was used in period houses, churches and other buildings and may last hundreds of years. Modern leadwork is generally the thinner milled lead which is more likely to suffer from 'creep' on slopes, flashings and other details, and from splits in exposed locations or where subject to movement. Generally a lifespan of 80 to 100 years is typical for modern leadwork. Modern lead is clearly preferable to mineral felt for flat areas and detailing, and the author prefers to specify leadwork where possible if the client's pocket is deep enough, especially in locations with a southerly aspect where mineral felt tends to rapid deterioration.

Large areas of leadwork without rolls or drips at suitable intervals may fail due to thermal movements which cause rippling and splitting. In its original state the lead is a bright silver colour but on contact with the atmosphere this changes to the familiar matt grey due to a coating of lead carbonate which forms on the surface. This surface carbonate protects the lead from further deterioration. The only chemical deterioration likely to occur is from strong acid attack and acid can occur in rainwater if this is discharging on to the lead from a roof area where there is moss growth, or material subject to moss such as ageing shingles. A point discharge such as regular dripping from a mossy roof will cause pitting to the leadwork due to such an attack.

Zinc was often used as an alternative to lead for flat roofs and detailing, especially in chimney stack flashings and soakers, and small flat bay roofs in traditional dwelling construction. In pollution-free localities the life of such zincwork is generally in the range of 20 to 40 years and sometimes even longer. Good detailing to zincwork is essential for durability, and the rolls and drips are necessary to accommodate thermal movements. When chemical pollutants are present in the atmosphere, the life of zinc is shorter and failure in the form of pitting and splits is possible after as little as 15 years. As with leadwork, a particular problem can arise with a constant drip of rainwater at a specific point over a period causing the acid in the rainwater to wear away the zinc and create a point for leakage. Zinc, like lead, also develops a carbonate on the surface which affords some measure of protection but this is thinner than the lead carbonate and does not provide as effective protection, especially where the atmosphere is polluted by sulphur dioxide.

Asphalt is often used for better quality flat roofs and is more durable than built-up felting provided that the surface decking is sound and free from excessive thermal movement. A lifespan of 20 to 40 years is generally found to be possible, depending upon the quality of the original job and the degree of exposure to sunlight. Deterioration is indicated by the appearance of surface splits and cracking, and by surface bubbles which form from the expansion of vapour trapped in the decking. Minor surface crazing may not of itself indicate general deterioration since the surface asphalt is trowelled into place and the trowelling is often overworked, as it is on stucco renderings which craze for similar reasons. This overtrowelling brings to the surface a skin of bitumen which may tend to craze due to slight differences in rates of expansion and contraction on the surface compared with the deeper material.

A normal thickness of asphalt is about 20 mm and anything less than this may be regarded as sub-standard and of a potentially shorter lifespan. Asphalt is not a suitable material for point loads, and will indent and spread under the weight of water tanks and other imposed weights. Very often considerable damage can be done to an asphalt roof by indentations created by chair legs, flower tubs, duckboards and the like. Most problems arise with the asphalt detailing. If the material is laid at angles to slopes and vertical upstands without suitable tilting fillets, creep will occur, especially in a south-facing location.

Copper is a traditional roof covering used for small areas of flat roofs for detailing, and for pitched construction in domes and spires where the distinctive green patina is required for aesthetic reasons. Copper sheet used for this purpose is thin but surprisingly durable, although not quite so durable as lead. In time the metal ages from the effects of thermal movements, and splits and ripples to the surface which will appear indicate that renewal is required. Rainwater with a high acid content attacks copper in the same way as with zinc and to a greater degree than in the case of lead. The thin metal – especially the rolls – is prone to damage from any foot traffic or ladders which may be used for maintenance.

Detailing

As a general rule all flat roofs should have a natural drainage fall and should not hold water. The reason for this is that ponding on the surface, especially with asphalt and built-up felt, will result in marked differences in temperature between the wet-cold areas and the dry-warm areas, especially in warm sunlight following overnight rain. This causes relative thermal movements between cold and warm parts which result in rapid deterioration of the surfaces. Generally, in new construction a fall of at least 1-in-40 is desirable to allow for possible bedding down of the construction in later life and to accommodate any minor deflection in timber joists. In existing flat roofs where any bedding down and deflection has ceased, a shallower fall may be acceptable but this should not be so shallow that the surface holds water.

Experience tends to indicate that most problems with roof construction arise at the detailing rather than the main areas of covering. Rain penetration of flashings and soakers of stacks and fire walls or to upstands and flashings at the perimeter of flat roofs is common. On close examination many roofs will exhibit a history of repairs to such details, the repairs often being of a highly unsatisfactory nature. Unfortunately the average building owner or lessee will rarely carry out any close inspections him/ herself and in the event of a leak will rely upon the judgment of a roofing contractor. The eventual bill will be paid without knowing whether or not the work done was necessary, suitable or reasonably priced. Often it is only when the building is subjected to the scrutiny of a surveyor that the nature of the past repairs comes to light.

Chimney stacks and copings

Particular problems may arise with downward damp penetration through soft brickwork, and mortar in chimney stacks and parapets. The design of stacks and parapets is often poor and for the correct construction of parapet wall copings two conflicting requirements have to be met. First, copings should be suitably mounted on impervious damp-proof courses and should be adequately weathered with overhangs and throating. Secondly, copings should be firmly bonded in place with mortar joints which will not deteriorate and be built of materials which are sufficiently durable and resistant to frost damage. Since the damp-proof course beneath the coping is a point where adhesion is poor, the design of copings must be a compromise between these two requirements. All brickwork to copings and stacks above damp-proof courses where exposure is continuous should be in special rather than ordinary-quality brickwork since the latter will suffer frost damage in time. Similarly, tiles, coping stones and other finishes to copings must be of a suitably durable type for the conditions of exposure involved.

Fig. 6.2 Traditional lead flashings are fixed by plumbers, not roofing contractors, this being part of the plumber's trade.

When chimney stacks are redundant and clients are quite sure that they will not be required again, an opportunity arises to remove the stack to below the roof line and to tile or slate over. This is often the most suitable solution as an alternative to rebuilding where a stack is in poor condition. When stacks are disused but in adequate condition they may be retained provided that suitable ventilation is applied at the top and bottom of the flues. If an unventilated cap is used or no ventilation is provided at the base of a flue, weather penetration downwards coupled with the effects of condensation within will often result in deterioration to the parging and brickwork and brown damp stains appearing on chimney breasts inside.

Gutters

Generally, all good roof design should incorporate the provision of adequate guttering and rainwater pipes to carry all roof water away to a storm drainage system. It is important that stormwater be taken away from the perimeter of the building to prevent damp penetration through walls and to avoid subsoil erosion from continual discharge of roof water into soil adjoining the main walls.

Certain traditional roofs encountered in period construction will not have gutters or downpipes, particularly in the cases of thatch and traditional stone slab construction where a deep eaves overhang is provided instead. It is vital with this type of construction that the area around the building should be paved, with a suitable drainage fall away from the main walls and with a good clearance between exterior and interior levels. A rendered plinth treated periodically with a bituminous waterproofing paint would also be advised. An examination of a number of thatched cottages over the years will reveal that dampness at the base of the main walls, especially on the weather-sides, will invariably be a problem unless the run-off of water from the roof slopes is adequately catered for. A combination of thatched roof, solid rendered walls of stone and lime mortar, and high exterior ground levels is often encountered; not surprisingly in such circumstances, invariably there will be a damp problem at the base of walls internally with a history of plaster repairs, or false walls applied over damp plaster.

A client proposing to buy a cottage with a thatched roof may have seen it through rose-coloured spectacles, especially if he first viewed in summertime. Such a client may have thought it quaint that visitors step down into the building instead of up, and will view uncritically the moss-covered and loosening rendering around the base of the building. The vendor's panelling to dado height around the ground floor rooms could be considered part of the charm. It may not be until the following winter that the client begins to detect the

Fig. 6.3 Modern gang-nailed trussed rafter roofs require diagonal bracing and gable-end strapping to prevent racking.

Fig. 6.4 Hampshire thatch with eyebrow window details.

effects of damp penetration at the base of the walls internally and that the carpet underlay is becoming mouldy around the skirtings. Perhaps then the significance of the high exterior levels and the nature of the construction will be appreciated and the client may feel inclined to look out the surveyor's report to see what was said on these subjects.

Often, small projections devoid of gutters or rainwater pipes are built and this will be commonly encountered in small bay roofs and porches of all ages. No matter how small the projection, guttering and downpipes are always advisable unless the roof overhang is very deep. Where these are omitted it will often be found that exposed sills and woodwork beneath have deteriorated from the effects of dripping rainwater. Staining down walls is also commonly found where guttering is omitted and this can cause considerable discolouration on north sides where moss growth is then encouraged.

Rainwater will occasionally be found taken to water butts instead of to stormwater gullies and this may be unsatisfactory, especially if the roof area is large, since overflowing from the butt will be undesirable. If water butts are needed it is suggested that gullies also be provided to take surplus water from the butts.

Modern PVCu rainwater systems have been found durable and maintenance-free if correctly installed with adequate brackets and sponge seals. Older rainwater systems will be encountered which may be less durable or suffer other drawbacks. Cast iron can be long-lasting if properly maintained but maintenance is often expensive and consequently poor and the hidden surfaces such as gutter interiors are often left unpainted. Asbestos-cement rainwater systems appear to have a useful lifespan of about 30 years and then become porous. Pre-cast concrete 'Finlock' gutters, commonly used in the nineteen-fifties, need periodic internal bituminising to avoid downward penetration of water into the walls beneath, especially at joints. Where an existing rainwater system needs repair and repainting it is often advisable to consider complete renewal in PVCu since the cost is often comparable and the final result should require no further maintenance in the foreseeable future.

7

Joinery and woodwork

> The woodworm is some of the worst I have seen. The only thing that is holding the roof together now is the fact that that the grubs are holding hands in the holes.
>
> *Surveyor to worried client*

In 1990, I gave evidence to an enquiry by the Health and Safety Executive (HSE). And following widespread consultation within industry and the professions concerned the HSE published a report in 1991 – Remedial Timber Treatment in Buildings, A Guide to Good Practice and the Safe Use of Wood Preservatives.

In his foreword to the report the then Parliamentary Undersecretary of State for the Environment, Tony Baldry MP, said 'This guide is important. It should be read by every surveyor or other professional involved in a decision to treat insect or fungal problems with wood preservatives.'

He went on to emphasise that whilst wood preservatives have a valuable role to play in maintaining and renovating our buildings by eradicating the insects and fungi that can damage and destroy structural timbers these chemicals are legally classified as pesticides and are, by definition, designed to kill living organisms. If they are not used properly they have the potential to damage both human health and the environment. It is, therefore, Government policy that the amounts of all pesticides used should be limited to the minimum necessary for effective pest control compatible with the protection of human health and the environment.

Tony Baldry emphasised that surveyors in particular take on an important responsibility whenever they decide that wood preservatives should be used. And in the guide it will be seen that in most circumstances the need for chemical timber treatments can be reduced or eliminated altogether by good design and proper maintenance.

Copies of Remedial Timber Treatment in Buildings can be obtained from Health and Safety Executive, Library and Information Services, Baynards House, 1 Chepstow Place, Westbourne Grove, London W2 4TF.

Wood-boring insects

The common furniture beetle (*Anobium punctatum*) is probably present in about 80% of pre-war houses in the London area and no doubt in a higher proportion of pre-1914 buildings. It is now becoming evident even in timbers in the earlier post-war houses dating from the 1950s.

Evidence of activity will be small circular flight holes, smooth inside and accompanied by deposits of bore dust – generally to be found in softwood roof timbers, floor timbers and staircases and, especially prevalent in plywood panelling under stairs and to joists and sleeper plates in suspended timber ground floors of inter-war origin.

The furniture beetle normally attacks only sound wood, although it is encouraged by timber which has been affected by damp. The flight holes are generally about 1.5 mm in diameter and the bore dust contains small faecal pellets which give it its characteristic gritty feel.

The deathwatch beetle (*Xestobium rufovillosum*) is often present in old hardwood beams and other hardwood timbers in period structures. On occasion, fungal attack may be required for this insect to become active so that damp conditions and poor ventilation would normally be an initiate for attack. Mostly oak is affected and many oak beams used in period buildings were formerly used elsewhere, either in other buildings or as ships' timbers, so that they were already second-hand, perhaps carrying with them deathwatch beetle grubs when first used.

Deathwatch beetle flight holes are larger than those of the common furniture beetle, being 2 to 3 mm in diameter. The bore dust contains characteristic ovoid pellets.

Wood-boring weevils (*Pentarthrum huttoni* and *Euophryum confine*) infest damp and decayed wood where fungal attack is also present. The flight holes are smaller than those of the common furniture beetle, being less than 1 mm in diameter. Weevil attack will commonly be found in the ends of joists and sleeper plates under suspended timber floors, and in other situations where there is contact between the timbers and damp masonry. Weevil attack often causes the complete disintegration of a damp sleeper plate or joist end resulting in the partial, if not total, removal of support from a floor.

Bark borer (*Ernobius mollis*) attacks the bark and sapwood of softwood timbers and is often found in roof rafters and joists. The flight holes and bore dust are similar to those of the common furniture beetle but if bark borer is confirmed, the surveyor will be able to reassure his client that no specific remedial action is required since this insect is not structurally destructive and will eventually depart of its own accord from the areas of infestation.

The powder post beetle (*Lyctus* family) attacks hardwoods, mainly oak, ash and elm, and especially elm. The insects are active only in the sapwood with the grubs reducing it to a fine powder. The activity of the powder post beetle may be distinguished from that of the common furniture beetle by the very fine, as compared with gritty, bore dust and the fact that activity is confined to the sapwood edges of roof timbers, boarding and other members. The elimination of powder post beetle requires the careful selection of timbers to exclude those with excessive sapwood before use.

The house longhorn beetle (*Hylotrupes bajulus*) attacks softwood timbers in the timber yard and in buildings. Because it causes rapid deterioration in the affected wood this pest is notifiable in certain areas. The flight holes are oval-shaped, 3 to 9 mm in diameter. A feature of the damage caused is that the timbers are eaten away internally, often with little external evidence to show the severity of the attack. At present this insect is generally found to be confined to buildings in the Home Counties but there are reports of it appearing further afield.

For information regarding damage caused, distribution within the UK, and the area within which preservative treatment of softwood roofing timbers is mandatory for house longhorn beetle, see 'House Longhorn Beetle Survey', BRE Information Paper IP 12/82, July 1982.

The larvae of the digger wasp has been known to attack teak timbers, normally thought to be immune from such infestation, resulting in almost complete disintegration.

Various other wood-boring insects will leave flight holes in new wood, especially imported wood, and may be found in new buildings where such wood is used. Pinhole borers may be occasionally found to have been active. However, some flight holes in new wood are permitted by current Codes of Practice since it would be unrealistic to discard useful timber merely because of a small number of old flight holes. In such circumstances it is sufficient to confirm that the timbers have in fact been suitably pre-treated before use and that all insect activity has ceased. CP 112, Part 2:1971, Appendix A.12 states that 'Scattered pinholes and small occasional wormholes are permissible in all grades. All pieces, however, showing active infestation should be rejected or subjected to preservative treatment in accordance with CP 98, Preservative Treatment for Constructional Timber' (superseded by BS 5268, Part 5:1977).

Wood-rotting fungi

Wet-rot is normally associated with the fungus *Coniophora puteana* but there are several other fungi capable of causing the symptoms of wet-rot and identification of the actual fungus responsible will be difficult until an attack is at a severe and late stage.

The main factor distinguishing a wet-rot attack from the true dry-rot (*Serpula lacrymans*) is that wet-rot requires damp timber in order to become active and spread and will confine itself to the affected and immediately adjacent timber. Dry-rot requires a humid atmosphere to become established, and damp timbers in situations lacking ventilation are ideal targets for attack. Once established, however, dry-rot fungus will extend readily into adjoining sound timbers and will travel through masonry and plasterwork in order to reach other sound timber which then in turn becomes infected.

Wet-rot is most often found in external joinery, especially lower window frames and sills, doorposts and wood thresholds in contact with the ground. Garage doorposts and fascias subject to driving rain penetration can also be affected. It is also found in roof timbers around stacks where weather has penetrated and in cellars and sub-floor areas subject to dampness. In an advanced stage the fungus will produce dark brown or black strands similar to old cobwebs and these will often be found forming patterns on the wall plaster in damp cellars.

Wet-rot is very common in the painted softwood window units used in the 1950s and 1960s due to the lack of adequate treatment of such timber before use. Often complete replacement of such window joinery will be found to be necessary due to this problem. In coastal areas and areas of high exposure this defect is quite general. In certain districts reference may be made to 'south coast rot' which is essentially due to difficulties arising from exterior woodwork of poor quality used in post-war speculative developments. One avenue of investigation open to the surveyor during his or her inspection is to probe carefully the exterior woodwork, paying particular attention to the obvious areas at risk and including especially any porches, conservatories or similar timber-framed structures, particularly those facing the prevailing weather.

Remedial measures in the event of a wet-rot attack in exterior woodwork should be made by cutting out affected wood well back from the point of decay and splicing in new pre-treated timber, preferably also using pre-treated wooden dowels and connecting with suitable tenons. Considerable care is required to achieve a proper repair and if wet-rot is extensive, this is often not justified. When rot is widespread, complete replacement with new units will generally be preferred. Filling repairs to exterior woodwork affected by rot will often be encountered in practice and such repairs must always be regarded as unsuitable as the decay will invariably return. Filling repairs, painted over, could be used by a vendor to disguise the nature of the rot and with a view to helping the sale of the property.

When the property under review is suffering from attack by the true dry-rot fungus, the necessary remedial action will include not only the replacement of all affected wood with new pre-treated timber but also a suitable programme of treatment for adjoining timbers, brickwork, plasterwork and sub-floor areas well away from the actual point of decay. Considerable care is required in devising a suitable specification for such works to ensure that the fungus is eradicated.

Generally a dry-rot fungus will be more likely to attack sub-floor timbers, spread unnoticed behind panelling and inside flat roof voids, all being points where a warm damp atmosphere may arise due to poor construction and lack of ventilation. Such conditions will mostly be found in older buildings where the amount of timber built into the structure may be considerable. Timbers such as those in stud partitions, lintels and wall plates may provide a passage through the structure which the fungal mycelium can travel, sometimes for considerable distances. The dry-rot fungus is able to spread in this way and attack dry timber because of its ability to produce water-carrying strands, or rhizomorphs, which are modified hyphae and are vein-like structures which can be as large as 10 mm in diameter. These rhizomorphs convey water from damp wood into dryer wood elsewhere, passing over masonry and through mortar joints in walls. They also 'store' food so that even if the rotted wood is removed the rhizomorph is still able to grow and infect new wood.

The mycelium of the dry-rot fungus is a snowy white colour (as compared with the brown or black strands of the wet-rot) and grows rapidly in damp humid conditions. Where the edge of the mycelium is in contact with light it turns bright yellow. Timber decayed by the mycelium exhibits deep fissures and the wood breaks up into cubes with the cuboidal cracking becoming apparent on the surface. Eventually when an attack is well established the sporophore or fruiting bodies will appear, these being thin and pancake-like, white around the edges, giving off rust-coloured spores and a strong mushroomy odour.

The sporophore or fruiting bodies of the wet-rot fungus are rarely encountered indoors but may occasionally be found in decayed external fascias and sills. They are smaller than those of the dry-rot fungus, generally brown in colour and in the form of thin plates attached to the face of affected wood.

In modern structures, one particular area where timber decay from either wet- or dry-rot is possible is within the void to a flat roof, especially if this is of sub-standard construction. Built-up felt roofs will often be found without any void ventilation and lacking any effective vapour checks. The climate within such roofs is ideal for the rapid growth of these fungi. The decking material may be Stramit board or chipboard which conveys water from a small surface leak by absorption through the roof area without any staining becoming apparent to the ceilings beneath. In such circumstances the dampness in the roof void, coupled with the warmth from the sun on the roof surface, will produce very rapid deterioration so that eventually when the roof covering has to be stripped for renewal, renewal of the decking and joists may also be found to be required.

An inspection of timber members used within the structure of a building can rarely be complete even if the building is empty of all contents and floor coverings and exposure of floor joists in all areas is possible. The reason for this is that some timber surfaces may be permanently hidden and lie within the structure in such a way that precludes a full inspection of all surfaces without taking the structure apart.

Timber frame construction

A surveyor may be asked to prepare a report on a timber-framed building, that is a building in which the outer walls are constructed in load-bearing timber work instead of traditional brickwork, blockwork or stone. Construction time can be much shorter on site since the panels will generally be delivered in a partially assembled form. Also the standard of thermal insulation can be higher and the cost of the main structure lower than would be the case with traditional construction.

In 1996, the Building Research Establishment (BRE) published three reports confirming the long-term durability of timber frame dwellings built between 1920 and 1975 (Timber frame housing 1920–1975: inspection and assessment. Timber frame housing systems built in the UK 1920–1965. Timber frame housing systems built in the UK 1966–1975).

In these reports the BRE found that the incidence of timber decay in the dwellings inspected was generally very low and with a few exceptions any decay was localised, superficial and usually the result of poor workmanship or inadequate maintenance. The BRE concluded that the performance of timber frame dwellings built between 1920 and 1975 is generally similar to that of traditionally built dwellings of the same age. The two reports on systems provide detailed guidance on the identification, construction and performance of 54 individual timber frame housing systems. The BRE concluded that, provided that regular maintenance is carried out and repair and rehabilitation work meets accepted levels of good practice in design and construction, a performance comparable with traditionally constructed dwellings of the same period should be maintained into the foreseeable future.

The survey of an existing timber-framed building presents certain difficulties as all important structural elements are hidden. There are nine main considerations in good timber frame design which are:

1. The design of the main structural members and fixings to resist all forces to which the building will be subject.

2. The adequate pre-treatment of timbers against rot and beetle attack.
3. The provision of damp-proof courses and membranes.
4. The provision of vapour checks in the wall panels to prevent interstitial condensation.
5. The provision of barriers against weather penetration.
6. The provision of adequate external claddings.
7. The provision of adequate thermal insulation within the panels.
8. Breather membranes in the walls to allow trapped moisture to escape.
9. Fire checks in the cavities of connected timber frame buildings to prevent the rapid spread of fire within the structure.

Fig. 7.1 Timber frame house under construction in an estate. Note the breather membrane over the timber frame inner skin to the prefabricated timber frame construction.

None of the foregoing matters can be checked from a superficial examination of the wall surfaces except item 6. A surveyor inspecting a timber-framed building may therefore have to discuss with the client the need for a structural check on the design from details available from the manufacturer and the possibility of opening-up selected points or using an endoscope to confirm hidden details. The local authority or other checking authority responsible for building control will normally insist that full structural calculations are submitted with deposited plans and that items 1 to 9 above are provided for before approval for a timber-framed building is given; details deposited with the local authority may be available for inspection but copies cannot normally be taken without the copyright holder's consent.

Suspected variations in the actual building as compared with approved plans should be checked, especially in the sizes and locations of openings in main walls or variations in the layout of the timber panels. This is most important as the structural principles involved in timber-framed construction are quite different from those of traditional brick, block or masonry building. Panels in timber-framed work require lining and diagonal bracing to give lateral restraint, unlike traditional work which obtains lateral restraint from the cross-bonding and intersection of walls. If lateral restraint in timber-framed construction is inadequate then the structure may be distorted parallelogrammatically by a process known as 'racking'. The danger of racking damage to a timber-framed building is at its greatest where openings are formed in the walls near to corners or if openings larger than specified are formed, thus reducing the overall stiffness of the framework. Detached timber frame buildings in areas subject to high wind loadings are most at risk.

The surveyor could ask to see structural design details and calculations for any modern timber-framed building which he or she is asked to survey. Clearly it would be advisable for the owners of such buildings to keep a copy of the details and calculations to be passed on to subsequent owners since questions are bound to arise from time to time when such properties change hands. In cases of doubt it is recommended that the details and calculations be referred to a consulting engineer for checking, and in any event a careful comparison should be made between the design and the actual building with a view to detecting any changes which may be structurally significant.

Structural alterations or conversions made to an existing timber-framed building should be undertaken only with the benefit of expert professional advice and if such a structure is found to have been modified subsequently, the nature of any modifications should be carefully checked.

The external brick or block skin used as a cladding to modern timber frame designs is not of itself structurally important. It often comes as a surprise to clients to learn that a building which appears superficially to be built of brickwork is in fact of timber construction. In some designs, the base of the brickwork is used to hold down the base of the adjoining panels using angle clips but brickwork or blockwork claddings are otherwise not normally structurally important to the design.

Preservation

It will be appreciated from the previous descriptions of the main wood-boring insects and wood-rotting fungi that good timber preservation in the form of pre-treatment is necessary to ensure durability in new work. A surveyor's knowledge of preservation techniques is important to assess the defects which will be found in older buildings and the remedies which may be proposed.

Different species of wood show great variation in their resistance to decay and beetle attack. Generally softwoods are most vulnerable; for example, ash and beech decay rapidly if conditions are suitable. Hardwoods are more durable in most cases and may be resistant to fungal attack even in conditions where fungal activity would be likely. Oaks and teaks, for example, are remarkably durable in damp conditions. In all timbers the heartwood is most resistant to decay and the sapwood least resistant. However, the sapwood will more easily absorb preservatives so that preserved sapwood of a non-durable species can be more durable than unpreserved heartwood.

Preservation is partly a question of good design to ensure that all timber members are kept dry and are not in contact with damp oversites or masonry. The climate around timber members should be cool, well ventilated and free from excessive water vapour. Areas most at risk will be suspended timber ground floors and enclosed roof voids, especially under flat roofs. In timber-framed construction the wall panels also require careful design, especially to give protection against weather penetration, provide vapour checks and allow entrapped moisture to escape through breather membranes.

Preservation also covers the treatments and surface coatings which may be applied to timber. Any water-proofing surface applied to timber will assist in preventing fungal decay since this will present a barrier to the absorption of moisture. Exterior woodwork is normally then painted in an oil-based paint, generally with two coats plus primer. Well-painted exterior surfaces are protected from moisture absorption by the paint and this degree of protection will be good if joints and putties are carefully sealed and painted over in such a way that no small passages for moisture remain. Painted surfaces reflect sunlight but surfaces with a hard clear finish such as varnish absorb the sunlight to some extent and this generally causes the surface to crack and peel quickly, especially on south-facing elevations.

An alternative to painting or varnishing is to give the exterior woodwork a periodic preservative stain finish so that instead of covering the wood with a surface weather-resistant shell, the treatment consists of regular brushing or spraying with absorbent preservative which soaks into the wood. It is possible that exterior woodwork treated periodically with an absorbent preservative stain will become more common in the future as the cost of traditional painting continues to rise.

To prevent fungicidal activity a wood preservative must so permeate the wood that fungal hyphae are unable to feed on it. To prevent insect activity within the timbers a wood preservative requires insecticidal defence properties against that specific species for which protection is required. Different wood-boring insects have different susceptibilities. Different treatments may be required for timbers where insects are known to be active and timbers where pre-treatment against future activity is required. In the former case eggs, grubs, pupae and adult insects may all be present in the wood and require eradication; in the latter case the protection required is protection against the ingress of egg-laying adults.

Chemical treatments should take a permanent or semi-permanent form to protect the timbers from fungal and insect attack for a suitably long period. Generally 30 years' protection would be a good standard at which to aim. This is possible because the chemicals are absorbed by and retained within the wood to a considerable depth and for long periods, unaffected by sunlight or other causes of deterioration.

Timber preservation may be undertaken in any one of five ways using chemical preservatives, these being: (a) pressure-impregnation, (b) dipping, (c) steeping, (d) spraying and (e) brushing.

Pressure-impregnation involves the use of sealed tanks of preservative into which timbers are placed, the wood preservative then being forced into the timber under pressure. This is suitable only for pre-treatment.

Fig. 7.2 A traditional brick and block house under construction in the same estate as
the house in Fig. 7.2.

Dipping involves immersing timbers in an open tank of preservative for short periods, perhaps a few minutes, and then allowing the excess preservative to drip from the wood.

Steeping involves treatment similar to dipping but extended over period of hours, days or even weeks, depending upon the nature of the timber involved and the degree of penetration required. Dipping and steeping are suitable only as pre-treatment before timbers are used.

Spraying may be used as treatment for *in situ* timbers. Generally a coarse spray nozzle is used so that droplet size is large and rapid saturation of the wood surface is obtained without producing a fine air-borne spray which could prove unpleasant or hazardous to operatives.

Brush treatments are often less effective than spray treatments since areas which are inaccessible to a brush may be reached using a spray nozzle, especially in suspended floors, behind panelling and in the extremities of roof voids.

Chemical wood preservatives are classified according to three main groups (and by reference to British Standard 1282:1959 (revised 1975)) as being tar-oil types, organic solvent types and water-borne types, and these are further sub-divided in accordance with the main chemical base of the preservative.

Treatment

Faced with an existing building which suffers from the effects of dry-rot, wet-rot or attack by wood-boring insects, a surveyor will have to make some recommendations as to a suitable course of treatment in addition to commenting upon mere facts and describing their significance.

It is suggested that the existence of wood-boring insects in timbers where the timber strength is not affected and where the activity is indicated by scattered flight holes only is best treated by spraying timbers until rejection, with an insecticidal preservative of a type appropriate for the species of wood-borer involved. The question of how far such treatment should extend into areas of adjoining timber construction is difficult, but in residential property, for example, it may be most sensible to confine treatment to the immediate area of infestation but at the same time to advise the client to have the other areas checked again after 5 years have elapsed. Thus, it may be possible to deal with the fact that flight holes indicate that insects have emerged to lay eggs elsewhere, the results of which could take some years to appear in the form of new flight holes. If timbers have been structurally weakened by woodborers it will, in addition, be necessary to recommend replacement or strengthening of weakened timbers, paying particular regard to the ends of floor joists and the stiffness of principal members in roofs.

A suitable specification for wet-rot attack may be prepared on the assumption that if the cause of damp can be removed and the timber allowed to dry out, no further attack will develop in timbers not already attacked. The specification should then include the renewal of all affected wood well back from the point of decay, the protection of surrounding timbers (necessary to prevent decay arising in the future) and the eradication of the source of dampness where this has been traced and found to have caused the attack. All new wood should be spliced and jointed into place to provide a lasting repair, free from weak points externally or inadequate weather protection. All replacement wood should be adequately pre-treated.

Specifications for dry-rot control require rather more details if these are to prove suitable, having regard to the particular nature of this fungus as already discussed. Depending upon the circumstances, a suitable specification should generally include the following specific provisions:

1. Cut out, carefully remove from site by the shortest practicable route and burn all the following: timbers showing cuboidal cracking, brown colouration, white mycelium or soft areas when probed; all sound timbers within 1 m from any of the foregoing defective timbers; all debris within roof voids, sub-floor voids or other areas adjoining, particularly timber off-cuts and similar debris.
2. Remove wall plaster, rendering, skirtings, wall panels, ceilings and other coverings as necessary to trace full extent of fungal activity in adjacent construction. Carefully clean out all exposed areas and

clean down all surfaces within a distance of 2 m from actual fungal activity. Sweep out all sub-floor voids, roof voids and similar areas. Remove all dust and debris from site by the shortest practicable route and burn.

3. Blow-lamp all masonry surfaces within an area of 2 m from area of fungal activity until surfaces are too hot to touch. When cooled, liberally apply fungicidal solution to all such surfaces.

4. Apply fungicidal solution to all remaining woodwork within 2 m from point of fungal activity in two liberal coats.

5. Make good to all timberwork in pre-treated wood which has been first given two good coats of fungicidal solution on all surfaces.

6. Make good disturbed wall and ceiling surfaces using 5 mm of floating coat of zinc oxychloride plaster over the rendering coat in the area of fungal attack and for a distance of 300 mm surrounding such area. Apply zinc oxychloride paint to painted surfaces where plaster is not to be applied and over a similar area and distance from previous attack.

7. Make good all design defects responsible for original dry-rot attack, including additional airbricks and improved ventilation, renewing defective damp-proof courses, preventing moisture penetration, eliminating condensation, clearing the bridging of damp-proof courses, repairing roof and plumbing leaks and timbers in contact with damp oversites or masonry as necessary and in accordance with good building practice.

Fig. 7.3 Flight holes of the common furniture beetle (*Anobium punctatum*) in relation to a £1 coin.

During a survey and when advising on timber treatment it is often helpful to be able to confirm the location and spacing of timbers behind surfaces in stud or similar construction. In addition to tapping the wall to check for hollow areas a useful tip is to use a compass which will indicate the location of steel nail heads and may enable the layout of timbers to be confirmed without exposure. Alternatively, metal detectors and stud sensors are now inexpensive and could form part of the survey equipment.

Finally, I would emphasise the importance of exercising prudence when advising clients to invest in chemical treatments. Some of the chemicals used now and in the past may be found to have damaging effects on the health of the operatives using them and the subsequent occupiers of the buildings treated. There is growing evidence to suggest that a great deal of unnecessary timber treatment may be carried out, much of it on the advice of surveyors and valuers who may find isolated evidence of furniture beetle or other infestations and refer clients to specialist contractors who will undertake extensive spray treatment using permethrin and – in the past – lindane or pentachlorophenol. We may expect to hear more about some of these chemicals in the years ahead as their long-term effects become better understood (see p. 225).

8

Finishes and surfaces

All of a sudden there was a noise like an express train running through the house. My wife opened the hall door and found that the whole of the landing ceiling had come down. The mess was unbelievable. Dust everywhere. Naturally we are holding you responsible.

Extract from tenant's letter to landlord

I thought that I should deal with all the parts of a building which a surveyor may look at, touch or tap in one chapter and to consider what potential defects or problems a surveyor may be looking for and what sort of advice a client might expect.

First and foremost in many buildings we will be looking at plasterwork to walls and ceilings. Increasingly now we will be concerned to discover if any asbestos has been used in the construction. And we will be looking at decorations. Also details such as wall and floor tiling. I think that it is fair to say that standards in all these respects have risen over the years and the present day purchaser is far more discerning.

I remember an old surveyor I worked with many years ago telling a client that the cracked and uneven lath and plaster ceiling would be best left alone and would be fine so long as he didn't shout too loudly. Nowadays, we have to say that if a smooth finish and durability is required then the client needs to budget for a new ceiling. Similarly old lath and plaster stud partitions with loose and live plaster which is extensively off-key have to be refaced.

Close examination of wall tiling is always a good idea because this can give useful clues as to the irregularities and past movements in walls and partitions.

With experience you can usually tell plasterboard, lath and plaster, fibreboard, asbestos–cement sheets and timber by tapping and the look of the surface. Plasterboard ceilings which have not been very recently decorated invariably show some signs of hairline cracking following the straight lines of the joints whilst with lath and plaster the cracking will be uneven and fairly random. Asbestos–cement sheet material used for ceilings will normally be battened at the joints because of the difficulty in sealing the joints using scrim in the usual way, so if you see a ceiling with the underside battened this is worth checking.

Plaster finishes

In most parts of the UK, dwellings built after 1938 will have ceilings of gypsum plasterboard. Exceptions to this will be found occasionally where timber laths and plaster or steel-mesh lathing and plaster have been used, but plasterboard will be the more normal. Prior to 1938, timber laths and plaster were generally used.

The life of the older lath-and-plaster ceilings will depend upon the quality of the original work and the degree of exposure to dampness from roof and plumbing leaks. Lath-and-plaster affected by serious leakage from any source will invariably have been weakened and will require renewal. During the course of the survey the condition of any lath-and-plaster ceilings should be checked by probing from below and, where possible, examination of the plaster key from above. If loose, soft or 'off-key' areas of ceiling plaster are found, general renewal in plasterboard should be recommended. Most lath-and-plaster ceilings will now have a limited remaining useful life and in most cases clients should be encouraged to budget for renewal in plasterboard.

Occasionally, fibreboard ceilings will be encountered, particularly in older buildings which may have been war-damaged during the Second World War, and also in some 1930s houses. This is inferior to plasterboard. It is not fire-resistant and is soft and easily damaged. Any chemical spraying of roof timbers for furniture beetle infestation undertaken over fibreboard ceilings will result in stains appearing beneath due to high absorbency, a point which should be borne in mind with this material.

Suspended ceilings of various types are commonly employed in commercial buildings and will occasionally be encountered in dwelling houses. In most cases there is no problem about lifting a panel or two to examine the void behind and this can be quite instructive as that void will probably be in its original undecorated condition with a past history of defects on view and services exposed.

Wall plasters in modern buildings should be checked for loose or shelling areas by tapping and probing as necessary. Problems of shelling wall plaster can often arise in speculative construction if walls are plastered when they contain a high water content and if the plaster used is almost impervious when set. Occasionally, major or complete replastering of new houses becomes necessary from this cause. The shelling and bulging of plaster finishing coats on clinker or breeze block walls and partitions generally arises about 6 to 9 months after completion. Replastering of the defective finishing coat can be difficult because of the high suction of the undercoat. The Building Research Establishment published a note on this topic in 1971, 'Shelling of Plaster Finishing Coats', BRE Information Sheet TIL 14, 1971.

Wall plasters which are well set after the first year or so in the life of the building should prove durable, but plasterwork in older buildings will become defective in the presence of dampness. Soft, damp and defective wall plaster in older buildings should be reported upon with the necessary advice. Where replastering is required, any special requirements should be mentioned, including the need for special plasters to be used where past dampness could give rise to efflorescence and surface staining of the new plaster.

Other wall and ceiling finishes

Walls and ceilings can have a wide variety of finishes other than plaster skim. Ceilings may be of plasterboard with scrimmed joints and may have a textured coat of Artex or similar finish. This will absorb the slight movements which concentrate in plasterboard ceilings and would otherwise cause a pattern of hairline cracks to the joint lines; it will not absorb more severe shrinkage cracks, however, and anything greater than hairline will probably appear eventually through the textured coating. Various patterns of swirls, combing and other effects are possible with textured coatings, depending upon the skill of the plasterer.

Walls can also be given a textured coating over existing plaster skim. In older construction, a stippled wall plaster may occasionally be encountered as an original feature. Whereas the modern textured coatings are slightly flexible, the older type of stippled plaster finishes do not have any flexibility and tend to be rather brittle. As a consequence, these stippled plasters to walls and ceilings may suffer cracking and the cracks may be difficult to eradicate or disguise since a stippled finish cannot be papered over with lining. If a smooth backing is required for wallpaper, the stippled surfaces have to be removed and the walls or ceilings re-skimmed in a new surface coat which could prove a major exercise in a house where all walls and ceilings are so treated.

Generally, plasterboard ceilings which are given scrimmed joints and a thin-board plaster finish will tend to suffer from a pattern of minor shrinkage cracks following the joint lines in the plasterboard panels. This arises more in ceilings to top floor rooms under timber joists due to the tendency for lightweight ceiling joists in roofs to be used for occasional maintenance access and storage which causes some flexing. If lightweight ceiling joists are used for an amateur loft conversion as a playroom or similar then the amount of flexing in the ceilings may be considerable and cracking to joint lines extensive.

A skim finish is not really sufficient for a plasterboard ceiling, and if crack lines are to be minimised then lining with a stout lining paper such as anaglypta or woodchip is recommended before painting.

All internal surface coatings which are applied in successive layers, such as textured coatings over existing plaster, are only as durable as the material beneath and the correctness of the application. A temptation arises to use surface coatings as a means of covering soft and loose plaster with the result that the surfaces eventually laminate or shell away from the bases. During the course of any survey it is recommended that wall and ceiling plaster be checked for loose and hollow areas and particular attention be given to areas recently covered over with new coatings or decorations in circumstances where soft or loose plaster beneath may be suspected.

Condensation

Problems of decorative finish to internal surfaces will be exacerbated by the presence of excessive condensation. Condensation is a major source of complaint in post-war buildings, especially buildings with a poor level of thermal insulation to external walls and poor overall design. Condensation problems at their worst can make a building uninhabitable due to the mould growth which forms on wall and ceiling surfaces and inside cold cupboards. This problem was virtually unknown in pre-war construction and has resulted from a combination of factors including changes in living patterns and construction methods.

Causes

In order to achieve a satisfactory surface finish, the effects of dampness and condensation must be dealt with at source. It is not sufficient to cover over problem areas which will subsequently reappear. Dampness to wall and ceiling surfaces may arise either from one or any combination of five possible causes, which are:

1. Direct weather penetration from the outside.
2. Leaking rainwater equipment and overflows causing a point of penetration.
3. Plumbing leaks within walls, behind ceilings or in adjoining areas.
4. Rising dampness originating in the site.
5. Condensation.

It is necessary to distinguish dampness created by 1 to 4 and involving specific remedies from condensation problems which are more complex and difficult to eradicate. Condensation occurs when damp air comes into contact with cold surfaces. Put at its simplest, the remedy lies in raising the temperatures of the wall and ceiling surfaces to above dew point. Alternatively, the amount of water vapour in the air may be reduced. Steps could be taken both to raise the surface temperatures and also reduce the amount of water vapour in the air. By such means condensation would be eliminated.

Remedies

Before dealing with the means by which surface temperatures can be raised it is necessary to discuss the more obvious causes of high water vapour content in the air. Certain simple precautions may be taken by the occupier of a building to reduce condensation levels by adjustments to living habits. Paraffin or flueless gas heaters discharging the waste products of combustion directly into the air within a building are a major source of condensation, and a change in the method of heating could be all that is required to eliminate the problem. Since the burning of hydrocarbons in oxygen produces large quantities of water vapour, paraffin heaters and flueless gas heaters of various types are really unsuitable in many situations and decorations are often affected, even on well insulated and warm wall and ceiling surfaces. Frequently windows will be found to be running with condensation where a paraffin heater or flueless gas heater has been in use for some time. One litre of paraffin produces about 1 l of water in combustion. Experience tends to indicate that this type of heating is used intermittently, often in older buildings and frequently by occupiers in lower-income groups for economy. Intermittent heating is itself a cause of problems since the air within the structure is warmed but the structure itself remains cold, an ideal condensation-forming situation.

Kitchens and bathrooms are obvious sources of water vapour, and when these rooms are in use their doors should be closed and good ventilation provided. The amount of steam in a bathroom can be reduced by the simple precaution of putting a little cold water in the bath first before running the hot water tap rather than vice versa. Appliances such as tumble dryers should always be provided with ducting to the outside air and damp washing should not be dried indoors in front of fires or over radiators. By means of small changes in heating methods and living habits, the amount of water vapour generated within a building may be substantially reduced and in a survey report a client could be advised accordingly.

Having dealt with the question of high water content in the air and the means by which it can be reduced, a condensation problem may still remain and this invariably arises from the bad design of the building, a matter about which a client will need to be informed. Poor design can arise from the use of unsuitable materials, from an illogical accommodation layout or from lack of attention to thermal insulation and ventilation.

The thermal insulation standard to apply as a suitable measure giving the performance of walls and ceilings should be that prescribed by current Building Regulations, and if that standard in a particular building falls short then it is recommended that this be explained to the client. Thermal insulation standards have risen appreciably in recent years so that many older buildings will fall well short of modern requirements. However, it should not be assumed that this fact will be accepted unquestioningly by clients. A client moving from a modern house with cavity walls and an insulation block inner skin, perhaps with a cavity filling and 100 mm of fibreglass quilt over first-floor ceilings, will need to be warned about the much lower standards which may apply in a house built inter-war with solid 225-mm brick walls and negligible ceiling insulation. In particular, the walls in bathrooms and kitchens in such a house may be prone to surface condensation, especially if the client keeps all the windows shut, takes frequent hot baths and showers and uses mobile gas heaters for heating. Such a client will not take kindly to the resulting peeling wallpaper and mould growth in the back of fitted wardrobes fixed to the external walls.

A note should be taken of the accommodation layout. For maximum efficiency, the greatest volume is enclosed by the least area of outside wall in a two-storey building in the shape of a cube (discounting curved walls). It follows that heat losses in single-storey buildings are higher and heat losses are also higher from projecting bays and wings. If a kitchen or bathroom forms a projecting wing to a building, perhaps on one floor only, then it may have three external walls plus a roof and a consequentially high level of heat loss. It will often be found that large old houses converted into flats will have small wings added to provide bathrooms or kitchens, or these may have been converted out of existing sculleries and outhouses with the original solid walls retained. In such circumstances these rooms may be cold, damp and ideal condensation traps.

The level of insulation above ceilings which are under flat roofs or balconies can rarely be checked without special exposure. In general, the heat losses associated with flat roofs and balconies are higher than would be the case where conventional pitched roofs are used. The interior of pitched roofs can normally be inspected and additional insulation may be provided without too much difficulty. The provision of additional insulation inside a flat roof is more difficult, necessitating either the lifting of the top surfaces or the removal of sections of ceiling. As a result, a number of methods have been devised for improving the thermal insulation performance of flat roofs, by the provision of either additional material above the decking or false ceilings beneath it, these then incorporating insulation.

In considering the design and thermal efficiency of a structure, small projections and areas of heat loss should not be overlooked. For example, small roofs to bays in older buildings often have no insulation at all. The floors of rooms located over integral garages, verandas and open porches should generally be separately insulated to reduce heat losses through those floors into the areas beneath. This is frequently overlooked.

Condensation is the likely cause if high moisture readings are obtained in wall plaster located on a cold external wall and readings are higher in the main and upper areas of the wall but lower at skirting level. Mould growth on wallpaper and within cupboards at high level is typical of condensation damage. Dampness from penetration is less general and related to specific exposed walls or points of penetration. Dampness from rising moisture will be concentrated at skirting level and will reduce in severity with wall height, rarely extending to higher levels above, say 1 m from damp-proof course level.

When mould growth on wall surfaces and inside cupboards is found this can be treated using fungicidal paints or fungicidal wallpaper adhesives. The use of fungicides to limit the development of moulds must, however, be regarded as a palliative rather than a cure and to be unsatisfactory in the long run. It is clearly preferable that the basic underlying causes of mould growths be dealt with by seeking out the sources of excessive water vapour or defective design leading to cold internal surfaces.

One useful means of raising surface insulation standards is to line the internal face of exterior walls with 3 mm polystyrene under wallpaper. This is particularly suitable in older residential property with solid walls and will help to compensate for the lack of a cavity. The disadvantage with this type of surface application is that the wallpaper and lining is prone to damage from denting. The overall thermal efficiency of the exterior walls is, however, much improved and condensation problems on walls are generally reduced or eliminated. When re-papering becomes necessary, the insulating lining usually has to be renewed at the same time.

Condensation on windows can be reduced appreciably by double glazing. Metal-framed windows are more prone to condensation difficulties than timber framed, particularly the older type of single-glazed galvanised-steel casements much used in pre-war and 1950s construction. Often the condensation from windows will cause staining to the reveals inside and this may be incorrectly diagnosed as weather penetration around the frame. The polystyrene lining already mentioned is effective in preventing such staining to window reveals but again, it must be regarded as an expedient rather than as a permanent cure.

Tiled surfaces

Tiled surfaces are usually found in bathrooms, shower rooms, cloakrooms and kitchens. Good quality wall tiling should be plumb, level with all joints, evenly proportioned and neatly grouted. It is a good idea carefully to examine wall tiling not only to inspect the standard of workmanship but also because the pattern of the tiles at wall junctions will give an indication of the evenness or otherwise of the wall surfaces and the extent to which walls may be out of plumb. Wall tiling, being inflexible, will show clearly the extent of any structural movement as accurately as any glass tell-tale so that gaps or cracks which have appeared in the tiling can indicate the extent of past structural deformation. Similarly, tile sills, internally or externally, may show signs of past separation at window openings.

The tiles should be tapped lightly all over to detect loose or hollow areas. The modern practice of tiling using spot adhesive is not always as durable as the traditional tiling applied to rendering encountered in older buildings. Older tiling will often be found to be very firmly fixed indeed and removal of tiles presents a problem. During renovation work it is often easier to re-tile over the original tiles rather than persevere with attempts to hack them off, and tiling over tiles can be satisfactory provided that the older surface glazing is cleaned and abraded first to provide an adequate key.

Tiling of various types is commonly used on floors. Over timber floors, the tiling should be capable of accepting flexing inherent in timber construction, while brittle tiles such as ceramic types, which require a firm solid base, are unsuitable for suspended timber floors. Vinyl tiles or vinyl sheet flooring can be laid over timber floors provided that an intervening layer of hardboard or plywood is provided and the degree of movement in the floor construction is minimal.

Exterior finishes

Exterior surfaces are often rendered and finished in pebbledash, spardash, smooth stucco, roughcast or stippled surfaces of various types. Such surface treatments must be suitable for the conditions of exposure in order to resist frost damage. They must also be designed to accommodate any movement in the construction beneath and to allow for the evaporation of trapped vapour.

In all cases the life expectancy of a rendered surface will only be as good as the adhesion to the surface beneath. Frequently, renderings are applied to walls which are in poor condition as a means of disguising the defects beneath and the key will be inadequate. In due course, the rendering will shell away from the base and the first signs of this, to be detectable by the surveyor, will be surface cracking and raised areas which are hollow underneath.

Various types of surface coatings are produced for painting roughcast and stucco renderings. Sprayed surface coatings are heavily promoted nowadays with various claims about the life of the treatment. As with the renderings themselves, the surface coatings will only be as durable as the base to which they are applied and the key is vital. Sprayed coatings are often used as a means of binding together loose or soft rendered surfaces and will not prove durable in such situations. Sprayed coatings are also employed as a means of limiting damp penetration through solid walls by providing an impervious coating and can actually be counter-productive in such a situation as, by preventing the escape of trapped moisture in the wall, they may increase a tendency for moisture evaporation from the internal surfaces where the barrier to penetration is incomplete or there is rising damp in the wall.

Some housing is given a new external facing of artificial stone on existing wall surfaces to alter the appearance of the building and (according to the manufacturers) improve weather-resistance and insulation. If such stone re-facing is encountered, certain basic questions need to be considered regarding the strength of adhesion to the base, the suitability of the base and any need to accommodate movement. Generally well-keyed brickwork or stonework in sound condition will provide a suitable base for re-facing, especially if shot-fired ties are used at suitable intervals. It is questionable, however, whether or not the re-facing of brickwork or stone in such conditions is worthwhile, discounting any aesthetic considerations. If brickwork or stonework is in poor condition and if re-facing is used for this reason, then this must call into question the long-term degree of adhesion between the new face and the base.

The application of traditional wall renderings requires degrees of skill, especially in the case of stucco which needs a careful selection of mix so as to be neither too lean nor too rich in cement. In such work, careful application is needed with a minimum of working of the surface. If stucco is overworked, the resultant surface will contain more finer particles than the backing and slight differences in the rates of expansion and contraction between surface and base will result in the familiar patterns of hairline cracking so often encountered in such a finish which has not been decorated recently.

Rendered exterior walls which are subject to constant water saturation from driving rain, leaking gutters, overflows and other causes, will be at risk from frost damage. South-west walls usually face the prevailing weather and will be most at risk. In parts of Wales, Scotland and the West of England where the degree of exposure is higher, it may be that all walls will suffer saturation from driving rain at some time. The crystallising into ice of the water between the rendering and the base will cause that rendering to fall away, sometimes large areas at a time.

Risk, probability and asbestos

Asbestos is a general term used to describe the fibrous form of several naturally occurring minerals. In buildings there are three types which may be encountered by surveyors: white asbestos (chrysotile), brown asbestos (amosite) and blue asbestos (crocidolite). Asbestos has been used in construction and elsewhere since 1860 with world production peaking in 1977. UK imports peaked in 1973. Asbestos was popular with specifiers because it is very light, fireproof and durable and because it mixes easily with cement, plastics and other materials. Its fibrous nature makes it a good insulator.

Asbestos is not specified now and its use was subject to gradual voluntary and legal bans from 1969 onwards.

Breathing in asbestos fibres can result in the development of mesothelioma (a cancer of the lining around the lung and stomach) asbestosis (a scarring of the lung) or lung cancer. Current evidence suggests that the section of the population most at risk from these diseases are those involved with the maintenance of buildings and associated services and the current annual death rate in the UK is in the region of 2000 to 3000 per annum with about 25% of these building maintenance workers.

We have a problem, however, identifying the time and location when a particular individual was exposed to asbestos leading to one of these diseases because the illness may only become apparent many years after exposure and the nature of the building maintenance industry is such that people move from one job to another and from one building to another. So we do not know in any specific case whether it was triggered by one event or by long-term exposure to asbestos over many years on many sites. This in turn obviously makes it very difficult for an individual to make a compensation claim for mesothelioma (which can only be caused by exposure to asbestos) because whilst we know that asbestos was the culprit we do not know where or when he, or she, inhaled the fibres that did the damage.

Current evidence suggests that the resulting disease can occur between 15 and 60 years after exposure and that smoking increases the risks. Blue and brown asbestos are much more dangerous than white and some commentators suggest that white asbestos on its own is no more harmful than mineral fibre, however, since white asbestos was often mixed with blue or brown all asbestos containing materials should be considered to the potentially dangerous.

The danger only arises, however, if there are airborne fibres which can be breathed in to lodge in the lungs. Asbestos fibres in food and water are not a heath risk and you cannot absorb asbestos through the skin. White asbestos mixed with cement in the form of asbestos–cement material, for example, is not a health risk when in a stable condition. The fibres can only be released into the air if it becomes friable or if it is sanded, drilled or broken up.

In 1995, Prof. J. Peto published an epidemiological study in the *Lancet* which revealed that those in the following list of workers were most at risk: builders, building surveyors, computer installers, construction workers, demolition contractors, electricians, fire and burglar alarm installers, gas fitters, heating and ventilating engineers, painters and decorators, plasterers, plumbers, roofing contractors, shop fitters and telecom engineers.

It is worth noting that building surveyors appear in Prof. Peto's list.

If you know the date a building was constructed this can provide a useful insight into whether asbestos containing materials may be present and if so where. Some useful dates are as follows:

- Sprayed asbestos coatings ceased to be used in 1974.
- Electric storage radiators may have asbestos in them up to 1975.
- Asbestos insulating board production ceased in 1980.
- Asbestos was used in textured ceiling coatings (Artex, etc.) up to 1985.
- Blue and brown asbestos were banned in 1985.
- White was used in thermoplastic and other floor tiles and bitumen products until 1992.
- Use of all forms of asbestos containing materials was finally banned in 1999.

The Control of Asbestos at Work Regulations 2002 apply to all commercial buildings and the common parts to blocks of flats and these regulations require whoever is reasonable for the maintenance to have an asbestos survey carried out to determine if asbestos is present. If asbestos is present they should have proper policies in place for its management. The regulations do not apply to individual dwellings or the interior of specific flats.

Having policies in place does not mean that asbestos containing materials have to be removed. Indeed in many situations the act of removal may be likely to release far more fibres into the air than simply leaving the material in place and containing it with suitable warnings to maintenance staff.

When carrying out a Building Survey asbestos containing materials will frequently be encountered and clients may need advise or reassurance. So let's consider some typical examples based upon properties that I myself have surveyed.

Soffits: Typically post-war houses built between 1950 and 1975 had asbestos–cement board material used to close off the eaves and verges using white asbestos (but with a remote possibility that some blue or brown could also be in the material). Risk to the occupants probably zero. But a decorator might decide to sand the material down prior to painting. The decorator would need to follow appropriate health and safety procedures – doing this in the open air with a face mask. Better not to use a power tool for this and just very lightly rub the surface to remove old flaking paint only, then decorate. Should the soffits be removed? Probably not. Better to leave them in place. If removal is required a properly accredited asbestos removal contractor should be used and this would be expensive.

Pipe lagging: A large 1930s country house with the old gravity heating system pipework still in place. In the loft we find blue or brown fibrous asbestos lagging to the pipes with the material soft and loose. The surveyor should immediately terminate the loft inspection, close the hatch and advise that a specialist asbestos removal contractor be employed to carry out a full survey and risk assessment. This is likely to cost a lot of money. Probably a small risk to occupants but a significant risk to any tradesmen carrying out work in the loft until it has been dealt with.

Water tank: A 1950s house with asbestos–cement cold-water tank in the loft. The tank needs to be replaced. Risk to the occupants probably nil. But the tank may need to be broken to get it through the loft hatch. A specialist asbestos removal contractor will have to be used. A simple job becomes potentially quite expensive.

Somehow, surveyors have to strike the right balance between advising the client and not being unnecessarily alarmist. It would seem logical to suppose that the risk to a person ironing clothes using an ironing board which has a small panel of white asbestos–cement material on the end must be very slight indeed and that person is far more likely to be electrocuted by the iron. Equally the decorator painting an Artex ceiling is far more likely to die falling from his ladder than from the effects of any asbestos in the Artex. And what is the probability that asbestos fibres from a water tank could find their way into the shower and onto the towel and then be breathed in by the person taking the shower who would then go on, perhaps 50 years later, to develop asbestosis?

So in the Building Survey report we have to record the presence of asbestos containing materials where they are found, we have to explain to the clients what is the current best practice in dealing with the material, give the clients the option to take further advice if needed and let the clients decide how they wish to proceed.

Surveyors and others working in conditions where asbestos may be present or investigating possible asbestos containing materials must follow appropriate health and safety procedures and have received appropriate instruction. Special face masks such as the Moldex 8000 type with the necessary performance grade of filter must be used. Ordinary builders' face masks of the type used in conventionally dusty conditions do not have the necessary fine grade of filter to deal with the very small asbestos particles.

9
Services

There are a number of things we are not happy about. For one thing the toilets are flushing hot water.

New house purchaser

Goodbye to the loft tank

In a severe winter the insurance companies pay out large sums to polyholders claiming for water damage – claims mainly arising from the freezing of pipes and tanks in roof spaces and the consequent damage when bursts take place. The insurers may well pause to wonder whether the traditional British plumbing system is the most suitable that can be devised for our climate – and for modern living – in the 21st century. Men have walked on the moon. This is the age of miracle and wonder. Is it really necessary to have a cold-water cistern in the loft and a copper cylinder in the airing cupboard with all those pipes?

In 1861, Thomas Hawkesley – and Englishman – invented the world's first unvented plumbing system with all outlets directly off the main supply pipe at mains pressure. And under regulations which came into force in July 1987, unvented systems were finally approved for use in the country of his birth. Initially the combination boilers all came from the continent because the British manufacturers were not geared up to produce them but we have since caught up. We now see them everywhere especially in flats where space is limited. I have encountered many unsavoury items in water tanks which are located in lofts or mounted on flat roofs, including dead birds and I am often surprised that property owners can be unconcerned about water tanks that are not properly covered or lagged.

There was once a good technical argument for water authorities insisting on the use of vented systems with storage tanks; that with such systems the contents cannot be back siphoned into the mains with a consequent risk of contamination of water supplies. But is not an argument that holds much sway (or even much water) now.

Plumbing folklore has it that the real reason why each house was required to have a water storage tank in the loft stemmed from the fear of invasion in Napoleonic times; that if each house had a storage tank it would be difficult for the Napoleon to invade and force a quick surrender by cutting off supplies. Each Englishman would bravely defend his castle safe in the knowledge that he could eke out a couple of weeks of drinking water from the tank in the loft.

Be that as it may for many years, having devised a complicated system that needs regular attention and often goes wrong, the plumbers were naturally reluctant to change it for anything simpler or high-tec. Special skills and tools are needed to deal with lead, copper and iron pipework. A multiplicity of fittings are produced, often incompatible with one another and thus needing a range of adaptors. And in the 1970s the advent of metrication was a further opportunity to introduce another complication with the metric copper pipework being slightly different from the imperial pipework it replaced (and in cheaper, thinner gauge copper).

Unfortunately most surveyors have a tale or two to tell about the vagaries of our plumbing systems and the variable quality of many of the installations, especially in modern speculative developments. And the damage that can be caused by a water tank left to overflow inside a house over the Christmas and New Year holiday, whilst the owners are away sunning themselves in warmer climes, has to be seen to be believed.

I have myself encountered water closets (w.c.s) flushing hot water, header tanks in lofts full of boiling water (thus generating a major condensation problem on the roof timbers) and heating pipework embedded in floor screeds with joints only a push fit because the plumber had forgotten to sweat together the soldered fittings.

A further problem is that the pipework is often totally boxed in and tiled over so if leaks do occur bathrooms and kitchens have to be taken apart in order to locate the offending plumbing.

In my experience the neatness of the job and attention to detail is often the best indicator of a good service installation. So if the wiring is all neatly clipped with circuits labelled, water, heating and drainage pipework similarly, this all suggests that whoever did the work took a pride in the job. If there are loose unclipped wires under the stairs or in the loft, or untidy and excessively complicated pipework in airing cupboards and around boilers be placed on guard that this may be an amateur installation.

Some clients will be remarkably knowledgeable about services and others will be entirely ignorant and require considerable guidance, especially regarding the effectiveness of heating systems.

During the course of the survey services should be examined at all points where they are visible. Great care should be exercised not to assume that services which appear modern have been renewed in their entirety. Amateur workmanship is rife in the areas of electric wiring, plumbing and heating; nothing should be taken on trust.

Electricity

The Institute of Electrical Engineers (IEE) Wiring Regulations recommend that all installations be periodically inspected and tested. For domestic residential buildings the recommended frequency is every 10 years. Places of work are covered by the Electricity at Work Regulations 1989 which require periodic inspection and testing by law without stipulating precise frequencies. The signatories to electrical certificates should normally be one of the following:

1. Chartered Electrical Engineer
2. Contractor Member of the National Inspection Council for Electrical Installation Contractors (NICEIC) or the Electrical Contractors Association.

A surveyor undertaking a building survey will not be qualified to undertake any testing unless he, or she, is coincidentally also a member of one of these bodies. If the electrical installation is obviously old and the wiring clearly in need of renewal, no electrical test is necessary since a surveyor can advise the client to budget for complete re-wiring and obtain a quotation before buying.

In all other cases a test and electrical certificate would be advisable whenever a property changes hands, and clients should be advised to have any work done which is necessary to comply with current IEE wiring regulations, the latest of which was the 16th edition published in 1991 and which in October 1992 became British Standard (BS) 7671, 1992.

Modern polyvinyl chloride (PVC) sheathed cable, used generally since about 1958, has proved durable and so far the author has not encountered failures in PVC sheathed systems except where grossly overloaded. Tough rubber, vulcanised India-rubber (VIR) and lead-sheathed cable, used prior to about 1958, must all now be regarded as due for renewal. Rubber-sheathed cable has a safe lifespan of between 20 and 25 years, tending to become brittle after this time, especially on power circuits used regularly for high-wattage appliances, and is liable to attack by gnawing mice in the case of older properties. Lead-sheathed cable can be mistaken for grey PVC cable, especially in, say, a dark roof void, but if a scrape with an insulated screwdriver leaves a bright shining line this will confirm the presence of lead sheathing.

The surveyor should be able to discover and comment upon certain obvious defects in electrical systems, and the following points should be checked:

1. At the service head, the overall condition of the switchgear and cable; generally a modern residual current circuit breaker (RCCB) unit with all circuits neatly labelled is to be hoped for, but invariably in older buildings a multiplicity of iron-clad or wooden boxes will be found and replacement of these by a modern consumer unit can be recommended. If rubber- or lead-sheathed cables are found, an assumption may be made that re-wiring will be required and the client should be advised to budget for this pending receipt of an electrician's report. It will be appreciated that ring mains used since the Second World War may themselves need re-wiring if rubber cable were used (mostly this applies to installations prior to 1958). Clients often believe that a ring main installation indicates that the system is modern and will therefore not need attention, but this is not always so.

2. The visible power sockets, light switches, pendants and fixed electrical appliances should be examined. For safety reasons cord switches are required in bathrooms and shower rooms where water is present and within certain distances of the switch. Immersion heaters should generally be on a separate 15-amp circuit with double-pole switch and have an indicator light. Such heaters on plugs and sockets, or spurred off a ring main, may not be satisfactory. A check should be made for frayed cables to immersion heaters and other appliances. Nowadays lighting pendants are earthed but this has not always been so. If the pendants are not earthed this may not, of itself, justify action but a client's attention should be drawn to the point. Unswitched power sockets are permitted by regulations but switched sockets are preferable. If unswitched sockets are found, clients may be told that these could be replaced with double or single switched sockets if required. Fixed electric wall heaters should be

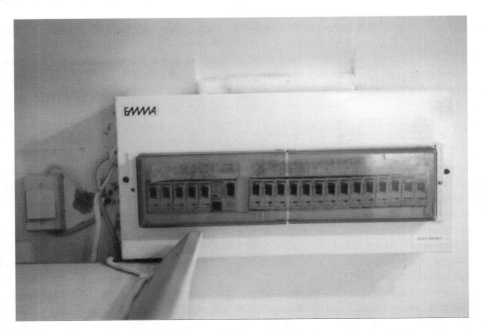

Fig. 9.1 RCCB unit with circuit breakers and adjacent bell transformer for simple domestic system.

on 15-amp circuits with fused isolating switches and in bathrooms or shower rooms should be cord operated and mounted high enough that they cannot be touched accidentally.

3. The visible cable on walls and in voids should be examined. Cable clipped to walls below a height of 1.5 m should be in conduit so as to be protected from damage. Unprotected surface-clipped cable run along skirtings is not satisfactory. In roof spaces the lighting cables should be taken to junction boxes, and although neat clipping is not essential it is desirable, particularly if the area is likely to be used for storage or occasional access. Wiring joined with insulation tape is unacceptable.

If the system is modern but obvious breaches of the Institution of Electrical Engineers' Regulations are found, or the installation has not been made in accordance with good practice, then the surveyor should be on his or her guard over the possibility of sub-standard or amateur work and advise a test before the system is used.

If the client wishes to cook with electricity, a separate circuit with 30-amp, 45-amp or 60-amp fuse should be recommended with a suitable grade of cable, depending upon the nature of the appliance to be used. A combined cooker panel and power socket is normally provided on the kitchen wall. It is essential that cookers are not run off ring mains, since this will result in overloading.

Any external circuits have to meet with special requirements and be suitably encased in armoured conduits, whether below ground or above. Exterior wiring is often amateur and, if so, may be dangerous. A particular point should be made of having any such wiring checked for safety.

Gas

A surveyor should recommend that the gas service, together with any fixed gas appliances included in the sale, be inspected and tested for safety as a normal routine precaution when the property changes hands and before any gas is used. The inspection and testing must be undertaken by a Council of Registered Gas Installers (CORGI) registered contractor. Having given this advice the surveyor may also make any other comments which result from his or her own inspection.

The author makes a point of examining the gas service head and meter, and if any gas smell is detected (a not uncommon occurrence) then this should be brought to the client's attention. If the building is old there may be a considerable amount of gas pipework running under the floor boards. It is not unusual to find original pipework for gas fires and gas lighting still in place and although this may be capped off where appliances were removed, it may still be connected to the service and at mains gas pressures. Older installations were designed for town gas which was not only supplied at lower pressures than modern natural gas but also contained more moisture, and this assisted in maintaining seals on threaded pipe joints. The dryer natural gas at higher pressures will often find weaknesses in such old pipework and leaks often result.

Occasionally, gas pipework will be found to be heavily rusted and for this reason it may be on the point of failing. Underground gas service pipework often leaks and requires renewal if old, especially if the ground in which it is laid is subject to vehicular traffic. The surveyor may mention this possibility and point out that any such underground pipework will not have been inspected as part of his or her survey.

A gas leak under the front garden of a house usually seems to occur just after some expensive re-paving and re-turfing of the front garden has been carried out – with adverse effects on the client's blood pressure!

Ventilation to conventional flued gas appliances is a matter of the utmost importance and should always be checked. Generally, for safe running a conventional flue boiler or warm air heat exchanger unit must have ventilation of the equivalent of two 225 mm $\times$ 150 mm vents to the outside air at all times. The increasing tendency towards improved insulation and draught-proofing of houses means that the structure becomes virtually airtight and if no ventilation for a conventional flue is provided, this could result in the suffocation of the occupants by removal of oxygen from the atmosphere.

A careful check of ventilation provision must be made and a warning given to the client not to alter or block ventilation in any way. Balanced flue appliances do not take oxygen from the room in which they are located and no requirement for special ventilation provision arises in this case.

Boiler location and other matters are governed by the Gas Safety (Installation and Use) (Amendment) Regulations 1996 which prohibit, amongst other things, the location of any flue under or adjoining any opening window. Examples of unsuitable flue locations will often be found during survey and should be mentioned to the client with suitable advice.

Certain small flueless gas room heaters have been available in the past and are still sold. This type of heater, like paraffin heaters, generates considerable vapour during combustion, and good ventilation for their safe use is essential. With this type of heating deterioration of decorations invariably results, even where there is good ventilation. If ventilation is poor and the fabric of the building cold, serious condensation problems occur leading to mould growth on walls and in cupboards.

Water

The plumbing for the hot and cold water supply must be examined where this is accessible. The location of both the company stopcock and the internal stopcock should be confirmed and if there is no sign of these the client must be advised to have them traced and checked.

Lead plumbing is still often found in older buildings and even if internal plumbing is modern, an original lead incoming and rising main supply pipe may have been retained. For health reasons lead pipes are no longer recommended. In the past it was believed that the lead could not enter the water in the system due to the protection inside the pipes provided by the coating which forms when the surface of the lead oxidises. This is no longer considered sufficient protection since the health risk which arises does so from a total long-term build-up of lead in the body from a variety of sources of which lead plumbing may be only one. Accordingly, while it would be foolish for the surveyor to be alarmist on this topic, if lead pipes are found or any lead-lined tanks, the appropriate advice must be to recommend their removal and the substitution of polythene, copper or similar modern materials.

During the course of the survey, the accessible plumbing and all sanitary ware, taps, cisterns and waste pipes are visually inspected and tested by normal operation. The surveyor should note any weeping or corroded joints, dripping taps needing new washers and noisy plumbing due to water hammer or other causes as well as running overflows and the consequent possibility, if header tank ball taps are found to operate correctly, of a leaking coil in an indirect cylinder.

Occasionally in an older building, the plumbing may have been turned off at the stopcock and the overall age and condition of the system may be such that re-charging the system with water would be ill-advised. In such circumstances it would be better not to charge up the system for tests but to advise the client generally that having regard to the obvious defects in the system, it should not be used until inspected by a plumber or perhaps renewed. Old stopcocks may be difficult to turn and can even seize or break off after they have been opened. Considerable damage could be caused then to a property if defective plumbing is charged up in circumstances where it may be difficult to turn off and drain it down again quickly.

It is a good idea to check the fixings to washbasins as these are often found to be loose. The caulking to bath edges is important and it is difficult to maintain a good seal with acrylic baths due to their flexibility. Caulking around showers and baths with shower attachments is essential and a watch should be kept for any points where a regular spillage of shower water may have taken place with possible adverse effects on timbers or ceilings beneath. Water-closet waste-pipe and flush-pipe seals are often found to be weeping at the back of the pan and any long-standing leak in this area may have resulted in deterioration to the floor or ceiling beneath. This often arises if the pan is not mounted exactly square to the cistern, or if modern press-fit joints have been wrongly used, or if the pan is mounted on an insufficiently rigid timber joist and boarded floor.

Certain combinations of metals in plumbing systems can give rise to galvanic corrosion. Galvanised-steel cisterns and hot tanks, for example, will rust through rapidly in the presence of a lead and copper piping. A sacrificial anode consisting of an aluminium block earthed lead to the steel tank may extend its life, although all galvanised-steel cisterns and hot tanks will rust through in time, whether protected or not, and replacement with a modern plastic or glass-fibre type is usually necessary once rusting has set in.

Central heating

Figure 9.2 illustrates diagrammatically the layout of a typical water- and central-heating system likely to be installed in a small residential or commercial property dating from the nineteen-sixties or nineteen-seventies. Such a system is simple with only a single pump on the heating circulation and very little to go wrong. Provided that the system is correctly installed and the radiators and pipework are protected from chemical corrosion by an inhibitor, such a system should prove reliable and serviceable for 20–30 years before major renewals or repairs are needed. Occasional servicing and attention to the boiler and pump may be expected, and a minimum of annual maintenance is recommended before each heating season.

Certain elementary principles apply in such a system. Overflows should be taken separately to the eaves and the header overflow should not be run into the main cistern. All expansion pipes should be of greater diameter than corresponding feed pipes and an overflow pipe diameter should be greater than that of the corresponding supply pipe. Cylinders must be indirect otherwise a mixing of circulation water and hot service water will occur. It is essential that this should not occur as the addition of fresh water into the circulation will add oxygen and cause rapid corrosion to steel radiators and fittings. This may happen

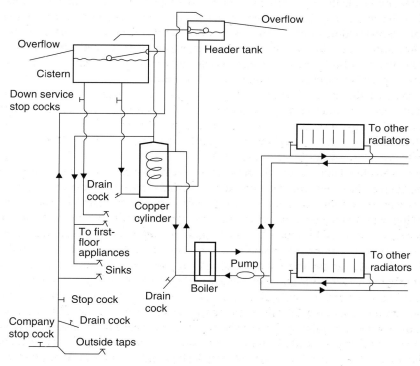

Fig. 9.2 Diagrammatic layout of a simple domestic water and heating system using copper piping in 15-, 22- and 28-mm diameters. Indirect gravity primary circulation is provided to the cylinder. A pumped heating system is provided in small-bore, two-pipe layout to pressed-steel panel radiators.

where the coil in an indirect cylinder has a split or pinhole and can be confirmed, as previously explained, by a continuous running of the system's overflow.

Occasionally it is puzzling to find an indirect system which does not have a separate header tank. This invariably will be a system with a self-priming cylinder, commonly used in the 1950s and 1960s, and can be confirmed if the cylinder has a 'primatic' label. Self-priming cylinders employ a one-way valve to permit the addition of water from hot into indirect service. In self-priming systems, defects can easily arise involving a mixing of the water and the lack of an expansion header tank means that inhibitors and corrosion proofing chemicals cannot be added to the system. A separate header tank and indirect cylinder are much to be preferred as seals have a habit of breaking down which can introduce rusty radiator water into the hot water service.

Some central heating systems are found to be laid out as a one-pipe rather than a two-pipe layout. In one-pipe systems, all circulating water enters and leaves every radiator via a single pipe, usually of wider diameter than normal small-bore or micro-bore. One-pipe systems can be gravity fed or pumped. In either case they suffer the disadvantage that the circulating hot water cools progressively as it runs through the radiators so that the radiators nearer the end of the circulation in a pipe run are colder than the earlier ones. Traditionally, the progressively cooler running can be compensated by employing correspondingly larger radiators towards the end of long pipe runs but this is of limited use when radiators are turned off. One-pipe systems are thus generally difficult to balance.

Gravity heating circulation uses wide-bore pipework of either copper or iron and is used in many older buildings including houses, blocks of flats and commercial structures of one kind or another. Gravity circulation is more sluggish than pumped circulation and older gravity systems inevitably contain a considerable sludge buildup with furring in pipework which further restricts the flow. A specialist heating engineer's report on older systems is inevitably necessary in order to confirm whether or not they are still functioning with reasonable efficiency.

All tanks and plumbing in lofts, under floors and at other points where frost damage is possible should be well lagged and tanks should always be covered. Various unsavoury items may be found in uncovered tanks including dead birds and, apart from the health hazard posed; there is a real danger of blockages occurring in the pipes and overflow outlet.

System control

In current domestic central-heating systems a system of control and layout may be adopted which is more sophisticated than the simple system illustrated in Figure 9.1.

A fully-pumped system could be employed using either two pumps, one for heating and one for hot water, or a single pump and two motorised valves, or a single pump and motorised valve giving heating or hot-water priority. The advantage of fully pumped systems is that the programmes for hot water and central heating may be quite separate, giving greater flexibility of control, and cylinder and radiator temperatures may be separately set using a strap-on thermostat to the hot water cylinder.

A secondary advantage of the fully-pumped system is that the boiler does not need to lie close to and below the level of the cylinder as is necessary for gravity circulation. The cylinder warm-up is also more rapid than with a gravity feed.

In addition to the use of a normal wall thermostat for setting room temperatures, systems of zone thermostats can be employed sited either floor-by-floor in a building or as thermostats controlling individual radiator valves. Generally speaking, the more sophisticated the system becomes, the more there is to go wrong and a point is reached beyond which the advantages of sophistication are outweighed by the disadvantages of complexity, including the cost and inconvenience of additional maintenance and repair.

Small-bore systems have, in some cases, given way to micro-bore, especially where central heating is installed in an older building previously not so heated. Micro-bore (or mini-bore) is easier to install since pipework is flexible and more easily concealed. Most micro-bore systems will incorporate some small-bore pipework around the boiler with the main pipe runs in micro-bore taken from central manifolds and either run under the floors or boxed in. Such systems require careful design to ensure that the small pipework is adequate for the velocity of circulation required to achieve the desired room temperatures.

Heat emissions from a boiler casing are equal to those from a small radiator so that if a boiler is located in a small room, say a kitchen or utility room, the benefit of heat emissions may be valuable in winter-time, making the overall performance more efficient. Often boilers are found to be located in exterior boiler houses, in which case the benefits of heat emissions from the boiler casing are lost on the outside air. In domestic installations the use of external boiler housing appears to be a practice dating from a time when fuel costs were low before energy costs began their present steep rise, and the overall efficiency of the system and prevention of heat loss were not regarded as being so important.

The controls for most systems incorporate timed switching. A programmer is preferred with a selection of timed programmes so that heating systems may be left to run automatically as required. In a residential property, a fully-pumped small-bore system with two motorised valves is ideal with a strap-on thermostat regulating cylinder temperature and a wall thermostat in the main living room regulating radiator temperature. Such a system may be either gas- or oil-fired. Under regulations which came into force in July 1987 continental-style unvented heated systems using combination boilers are now permitted in the UK. These systems have a pressurised boiler and there are no storage tanks.

In addition to conventional water-filled central-heating systems a wide variety of other methods of heating may be encountered which could from part of the structure under review, or alternatively are chattels, including various types of free-standing gas and electric fires, electric storage heaters, mobile gas and paraffin heaters and solid fuel appliances including wood-burning stoves. As such items are chattels included with the property it is recommended that clients be advised to satisfy themselves on the condition of appliances and their performance, including their running costs. Heating requirements are generally somewhat personal and a means of heating which will be satisfactory for one client may be unacceptable to another.

Flues

In the case of heating systems which use flues it is recommended that clients should be advised that flues must be swept and checked as a normal routine when a building changes hands. A buildup of soot in a flue can cause a chimney fire and this can be a hazard in an old building where a flue lining may be poor or non-existent. A risk can arise of timbers having been built into chimney walls, especially at roof level, this giving a combustible access for a chimney fire to spread into a roof space. Many cases have occurred in old buildings of beams or purlins having actually been built into a flue and smouldering for perhaps months before bursting into flame with disastrous results.

A cautionary note should be sounded over flues from wood-burning stoves which are often installed in existing fireplaces. Manufacturers of these appliances will often recommend that the dampers should be opened every 24 h and used at maximum heat to burn off the tar deposits which accumulate in the flue. The temperature of the flue linings can reach 850°C, and if tar ignites it could reach 1500°C, and serious fires have occurred as a result. Moreover, chemicals are sold to the owners of wood-burning stoves to assist in dissipating flue deposits. Chemicals using phosphorous cause intense heat and those including chlorine can result in chloride attack to linings, especially old parging and brickwork.

If roof and floor timbers are in contact with the walls of flues or even built into these walls (a not uncommon fault in older buildings), they are likely to be subject to intense heat to the point of combustion by this practice of flue tar burning. Only flue linings of a very high standard will resist such aggressive conditions and when a wood-burning stove is encountered, it is suggested that the suitability or otherwise of the flue arrangements should be reviewed critically.

Drainage

Stormwater

Stormwater drains generally consist of gutters and rainwater pipes which discharge into either open or sealed gullies and with surface-water gullies provided to paved areas. The stormwater may run to soakaways or to a separate stormwater sewer or into a combined foulwater and stormwater sewer. It is recommended that the surveyor should become familiar with the accepted practices in the locality concerned as local water authority regulations may provide, for example, that stormwater should not be run into the foul sewer, especially in inland locations where large quantities of stormwater would overload the foul system. Soakaways are suitable in porous, well-drained soils but unsuitable for heavy clay or non-porous soils which drain reluctantly, unless there is no alternative.

Frequently the surveyor will find the storm drainage system to be a mystery. There will often be no manholes nor any indication of the drain runs. Soakaways tend to block eventually and the pipework becomes choked, especially if leaves fall in the gutters and no cages are provided over rainwater pipes. Soakaway systems often need to be relaid after a period of time as they cannot easily be rodded through.

The importance of avoiding point discharges of stormwater into the soil around foundations has already been stressed in Chapter 3. In many cases the surveyor should consider whether or not the storm system ought to be exposed and checked. Perhaps the best guide to this is to suggest that a client checks the system during a period of really heavy rain to see if the storm gullies allow a free flow. If they are blocked it is likely that stormwater is draining into the soil around the building from overflowing collars and gullies, and this is undesirable. A good modern practice is to lay storm drains watertight for a distance of at least 3 m from the main walls and thereafter in land drains to a soakaway if required. Some intermediate rodding access is recommended in the form of small rodding chambers over any bends. If the storm drains run into the foulwater sewer or to a separate storm sewer, a watertight system of a standard similar to the foulwater system is required with suitable rodding access.

Foulwater

Foulwater drains should always be provided with adequate means of access for rodding and attending to blockages, and provision for rodding at each change in direction and each branch drain connection is necessary. If this is lacking, it is suggested that this be mentioned in the report together with suitable advice. Lack of good rodding access on a branch drain from a gully may not be critical but lack of access to a branch drain from a WC or soil pipe serving a WC is unacceptable if a good standard of design is to be maintained.

Foulwater drains will normally take the form of either a public sewer or a private sewer or a private drain. These distinctions are important since they affect the client's potential liability for drain repairs both within and outside the client's own curtilage.

Sewers

A public sewer is maintainable by the local water authority or its agents and will normally be laid under a public highway or grass verge but can also run under private land. Generally, defects in a public sewer will not be the liability of the adjoining property owners connected to it once it has been adopted and it is not normal practice during the course of a building survey to attempt any examination or testing of public sewers serving the property under review. In some localities the public sewers, both for stormwater and foulwater, may become overloaded from time-to-time, and if the surveyor is aware of any such problem it would, of course, be advisable to discuss it in the report. The ability of sewers to cope with storm- and foulwater discharges may be relevant when changes of use or more intensive use are proposed. Changes of use normally require local authority permission and it is possible for consent to be refused if the existing public sewers are inadequate. Such cases are, if anything, on the increase, particularly in urban areas.

In England and Wales a shared drain laid after 1937 may be a private sewer for which all property owners connected are jointly liable up to the connection with the public sewer. In order to be able to confirm the extent of a client's potential liability for a private sewer, it would be necessary to inspect and possibly test the whole system. Clearly this is often impracticable; frequently 20 or more houses will be connected to a private sewer and a surveyor could end up testing half the drains in a street in order to cover that potential liability. Where the subject property is connected to a private sewer, it is suggested that the significance of this be explained to the client and that he or she be given the opportunity to decide how far to pursue the matter.

In addition to the Public Health Acts which impose a liability for the maintenance of private sewers which can be enforced by the local authority, the question of responsibility for shared drains may also be covered in title covenants for freehold property, or lease covenants for leasehold property, or their Scottish equivalents. If the title or lease specifically identifies drains for which a client may be liable, these may require special attention from the surveyor, possibly needing access to other parts of a building or adjoining properties.

It is rare for a local authority to serve notices under the Public Health Acts requiring repairs to a private sewer but this situation does occasionally occur. It is clearly a matter of concern to a client buying a property to find that he or she may have to pay a contribution for some drain repair, perhaps some distance from the property and about which he or she may be unaware until served with a formal notice. It is debatable whether or not all property owners connected to a private sewer must contribute or only those 'upstream' from the defect, as local authority practice in serving notices under the Public Health Acts appears to vary.

A private drain is a system which is exclusive to a particular property and runs through its curtilage and into a public sewer. Up to the point of connection to the public sewer the private drain is entirely the responsibility of the property owner (or lessee) concerned. Testing procedures for private drains are clearly much simpler than for private sewers.

Testing and maintenance

Two main forms of drains test are normally used by a surveyor, these being water-pressure tests (particularly suitable for traditional pipework between open manholes) and smoke tests (particularly suitable for above-ground soil stack systems). Drain testing procedures and the various types of drainage systems are well documented in the building textbooks and are not discussed in detail in this volume. However, certain comments on the interpretation of test results would seem worth stating.

The first point to bear in mind is that the standard of test result required for new drainwork will rarely be achieved in existing drains. Old drains may well be watertight in normal use when running perhaps one-third full and, as such, may be perfectly serviceable. A water-pressure test to any height will impose abnormal pressures on the pipework and manhole benching and will often indicate leakages. Such leakages may not of themselves justify the relaying or pressure-grouting of the pipework, and the degree of leakage, the invert depths and the general condition of the system should be taken into consideration when deciding whether or not major repairs are justified.

It is an undoubted fact that a majority of pre-war drainage systems will leak under test if the head of water used is high enough, but many will flow freely and be otherwise quite serviceable under normal conditions.

It is perhaps worth considering that a client might expect from his drainage system:

1. That the drains will flow freely in normal use without blockage.
2. That in the event of blockage occurring it can readily be cleared.

3. That the system be airtight at ground level and within the building so that no drain smells or health hazards arise.

   Most common blockages are caused by stepped fractures in pipework, tree-root intrusion or rough workmanship to joints and manhole benching. These matters will often be detected on a visual examination within manholes and through pipework using a mirror. Shallow drains running under or close to trees should always be checked for tree-root intrusion, even if the test result is reasonable, since the fine root tails can enter the drains through narrow fissures and even effectively seal them off in some cases.

   If the drain gradient is too shallow this may cause blockages and rough checks on the invert depths will generally indicate whether or not the gradient is too shallow. If a level is not available, a check can often be made on drains around a building by taking invert depths in relation to the main damp-proof course line of adjoining walls (assuming damp-proof courses are not hidden).

   Manhole covers will, more often than not, need cleaning, greasing and re-setting to provide an airtight fit and cracked manhole covers should always be renewed to keep them airtight. Loose frames should be re-set in new cement haunching for the same reason.

   In some districts fresh-air inlets in conjunction with interceptor traps were used on private drains in the past but may no longer be regarded as necessary where more than one vent pipe is provided. When this is so, I generally suggest that fresh-air inlets be sealed airtight in cases where other ventilation is to a suitable modern standard. The exception is where an interceptor trap is fitted in the last manhole before the sewer and a fresh-air inlet with mica flap valve ventilates the drain in conjunction with a single vent pipe at the head. Mica flap valves in a fresh-air inlet to keep out vermin are often found to be broken or missing.

   Some modern unplasticised polyvinyl chloride (PVCu) underground drain systems, such as the Marley rodding point system, dispense with traditional manhole construction for most purposes and instead provide angled rodding eyes at ground level for each change of direction. Testing using drainplugs or inflatable bags is still possible but a visual examination is not possible through the pipework without recourse to video equipment and this is not usually suitable for the narrowest pipe diameters.

   In rural areas, private drainage systems encountered vary from the modern, efficient and self-cleansing to the ancient and insanitary. The author has encountered drains with no access whatsoever and no indication of where they terminate, also drains discharging directly into ditches and watercourses or into bore-holes. For example, in the Mendip Hills, house drainage systems used to discharge into potholes known locally as 'swallets' which eventually directed the effluent into the cave system beneath. Clearly such primitive systems are often highly unsuitable and advice may be given that a modern system with ample access for cleansing be provided instead.

Cesspits

Cesspits are watertight sealed tanks which are periodically emptied by the local authority, their agents or specialist contractors and subject to emptying charges. If possible, the capacity of the tank and the amount currently charged for emptying should be checked. Clients may then consider if the tank capacity is adequate and the arrangements for emptying suitable. It is a common practice for local authorities to insist on watertight cesspits and for these to be built in 225 mm brickwork, rendered both sides. The author has heard of occasions when empty milk bottles would be built furtively into the brickwork and rendered over so that when the cesspit had passed final inspection, the ends of the bottles would be broken out thus permitting the tank to drain into the surrounding subsoil and reducing the frequency and cost of emptying. Rustic cunning of this type might be expected if the cesspit is required to be emptied only at infrequent intervals. The local authority or their agents will generally keep a record of emptying dates for all cesspits in their area and a telephone call can often elicit details of the number of emptying over the last few years together with the amounts emptied.

It is essential that cesspits be located within range of the suction pipe from an emptying vehicle and that a parking area for the tanker is provided with suitable access from the highway. If there is any doubt about the correct location of the cesspit, this point should be checked with the local authority.

Septic tanks

Septic-tank systems are well described and documented in the textbooks and the author will omit any technical details of these. If a report is required on such a system, the following points should be borne in mind:

1. Septic tanks, like cesspits, also need emptying periodically. The frequency of emptying is, however, much less with a septic tank, perhaps annually or even every 2 years, depending very much on the design, capacity and the amount of solid material with which it has to cope and, consequently, the amount of humus which accumulates in the bottom of the tank. Accordingly highway access for a tanker with a parking area within suction range is again necessary.
2. It will often be found that filters and soakaways become choked after a time and land drains sometimes need to be relaid if the system is old. Any evidence of a tendency for the system to block or flood may indicate choked soakaways. This depends to some extent on the length of the soakaway pipes, the lie of the land and the draining capacity of the soil. Gravel or chalk soils drain very well whereas heavy clays drain badly and are not suitable for taking soakaway systems unless considerably extended.
3. If there is any doubt about the design of the system or its suitability for the client's needs, a specialist drainage engineer's report could be suggested and the client given the opportunity to commission such a report before he or she buys the property. Always lift the covers to the tank itself and look for the 'crust' on the surface which indicates that the septic tank is working correctly.

An improvement to the septic tank is the modern Klargester biodisk system, which uses an electric motor and rotating disks to hasten the biological breakdown of sewage and produces a treated effluent suitable for discharging into a suitable water course.

Drains generally

The Royal Institution of Chartered Surveyors (RICS) Practice Note 'Structural Surveys of Residential Property' recommends that the surveyor undertaking a structural survey should open all accessible manhole covers, record the routes taken by the pipework below ground and observe the drain flow; also that the means of disposal of foul water and surface water be confirmed (whether by main sewer or otherwise) and that storm drains be tested with water to identify blocked gullies, etc.

The Practice Note does not require that drains testing be undertaken as a standard or automatic procedure but that the surveyor should note features which might justify further investigation or tests. Such features might be evidence of past blockages, shallow drains running under paving which has cracked or subsided, drains running under trees, older traditional earthenware pipework running in shrinkable clays or drains with cracked channels and damaged benching within manholes. If it proves impossible to see between manholes using a mirror this may well indicate fractured or distorted pipework.

Pitch fibre pipes

Occasionally pitch fibre pipes may be encountered. Typically black in colour in contrast to the red earthenware or the orange PVCu normally seen, these were used for a short time in the post-war period during the nineteen-sixties but were found to suffer certain drawbacks. The pipes can deform under loading from above and were prone to damage from vermin since determined rats could gnaw through them.

If pitch fibre pipes are found these will generally need to be renewed. A pressure-grouting repair is not possible and the drain has to be relayed. This is potentially expensive for clients and unless the nature of the pipework is confirmed on survey, the problem may only become apparent when leaks or blockages occur and a closed circuit television (CCTV) survey is undertaken.

Closed circuit television

Specialist drainage repair contractors offer surveys to customers which usually include a test of pipework under pressure and a CCTV survey with video. The camera travels through the drain and the video will show stepped fractures, tree roots and other problems. The quality of the video pictures tends to be rather poor but experience suggests that this is a useful tool for identifying defects.

Fig. 9.3 When mains drainage is not available a modern alternative is the Klargester Biodisk septic tank drainage system. Effluent enters the system and is rapidly broken down by bacteria with the process hastened by the electric motor and rotating disks which increase the exposure to oxygen in the air. The treated effluent emerging from the system is clear water suitable for draining into soakaways or an adjoining water course. Consent to discharge effluent is required from the Environment Agency. Periodic de-sludging is necessary.

10
The report

The defendant failed to exercise the reasonable care and skill of the prudent surveyor which he had impliedly promised to exercise when he received instructions to survey the house. He committed a breach of contract with the plaintiff and provided him ultimately with a report which was so misleading as to be valueless.

Judge Kenneth Jones – Morgan *v.* Perry *(1973)*

A Building Survey report should always be made in writing and be prepared with great care so that nothing intended to be included is omitted by mistake. It should be printed neatly on good quality paper and bound with a reasonably durable cover since it is a document to which reference may need to be made on subsequent resale of the property, or for some other reason, some years into the future.

The report should be prepared in narrative style with either numbered paragraphs or many headings and sub-headings (or both) so that in the event of discussion between the surveyor and the client, by telephone or otherwise, each party may readily refer the other to the section concerned. Pages should be numbered so that if a page is missing from a report it will be obvious.

As I have argued in my introduction to this book the report should not be unnecessarily long and any surplus comment, padding and attempts at humour are best avoided. Few clients would be happy about paying a fee for anything other than facts and information on the basis of which they will be taking an important personal and financial decision – possibly the biggest financial decision in their whole lives.

At the conclusion of the report it should be electronically signed off, or physically signed, by the surveyor concerned giving his or her name and qualifications.

Tenses

In some surveyor's reports a practice is made of presenting all facts and description in the past tense. This can seem rather strange to the client. To say 'the main walls were of cavity construction' or 'the main roof was of flat construction surfaced in asphalt' can engender a note of uncertainty in the mind of the client who may feel that this suggests that the walls or roof may now, somehow, have changed.

On the other hand to say 'the asphalt roof surface is free from ponding, surface cracking or vapour blowing' could be unwise because there is a possibility that conditions (such as the weather) may have changed in the interval between the inspection and the signing-off of the report.

So I would suggest that the present tense be used for all matters of description. Thus the walls are cavity construction and the roof is flat. But that the past tense be used for all comments on condition. So there was no evidence of unusual movements or settlements to the main walls and there was no evidence of ponding on the flat roof, at the time of inspection.

Drafting the report

In Chapter 18, I have reproduced one of my old reports, with some changes to disguise the actual address of the property. I have done so somewhat reluctantly in the knowledge that many fellow surveyors will find lots to criticise and there is nothing surveyors enjoy more than finding faults in one another's reports.

In those parts of the report dealing with defects or potential future problems there should be an order of presentation consisting of three parts:

1. the investigations made,
2. the information obtained from those investigations,
3. the advice which follows from the information.

So, for example, we may say the following.

We inspected inside the main roof space using the trapdoor and pull-down ladder set into the landing ceiling. Our inspection was limited to some extent by the reduced headroom at the eaves, insulation material and stored items.

Where seen the ceilings are original lath and plaster construction consisting of a lime plaster set into Chestnut or Willow laths. The plaster was soft and extensively broken and loose. Ceiling surfaces below were extensively cracked under the old lining papers. The life of lath and plaster ceilings will depend to some extent on the quality of the original job and the degree of exposure to dampness and vibration. All the ceilings to the top floor rooms have now reached the end of their useful life.

We recommend that you allow in your budget for replacing or underlining ceilings in new plasterboard before the next redecoration. If you are in any doubt regarding the potential costs involved you should obtain a quotation from a reputable local plastering contractor prior to legal commitment to purchase. It is not always essential to hack down the old ceiling plaster. It may be possible to apply new plasterboard to the underside of the lath and plaster using longer plasterboard nails thus avoiding much of the dust and disruption involved in ceiling replacement.

Now, of course, the client may decide to do nothing about the ceilings. Clients often forget all about the advice in the surveyor's report once they have moved in. But if one of the ceilings falls down later and this client re-reads the report he, or she, can have no cause for complaint against the surveyor.

Direct on-site dictation

An alternative method of carrying out a Building Survey and preparing a report is to dictate the report direct onto a digital voice recorder, for downloading onto a computer system, whilst walking around the property. This method is not recommended (see *Watts and Another* v. *Morrow* in Chapter 19).

There are two reasons why direct on-site report dictation can cause problems for surveyors unfortunate enough to find themselves subject to a claim.

First there are no notes on file. The report is all that there is. There is nothing to record what the surveyor did, or did not, do other than what appears on the face of the report. In the event of a negligence claim it may be difficult for the surveyor to defend the report if it appears that something has been overlooked. Notes and sketches taken using a suitable field sheet will at least demonstrate methodology and confirm that the surveyor was faced with a prompt or checklist to consider the possibility of a particular defect.

Secondly, there is no time for what the judge in *Watts* v. *Morrow* described as 'reflective thought'.

For example, the surveyor may see a damp patch on a wall and a test using a moisture meter may show a high reading. Since damp can be caused by moisture rising, penetrating or descending and might result from problems with the roof, gutters, pointing, damp-proof courses, bridged cavities, condensation or plumbing, a number of other areas of inspection have to be drawn together in order to reach a conclusion and offer advice to the client.

I take my notes in the form of sketches with simple line drawings at each floor level indicating door and window openings, chimney breasts and other points of interest. A simple sketch of each elevation will show gutters, downpipes, bridged damp-proof course and other issues.

Back at the office the damp patch inside can be seen in relation to the solid brick external wall and cracked collar to an adjacent rain water pipe. So an appropriate section of the report can be dictated following the recommended rules; first the investigations made, secondly the defect identified and finally the action the client needs to take.

Standard paragraphs and survey management software

Nowadays virtually all surveyor's reports are prepared using Microsoft Word or similar software as part of a computer-based survey management system and this speeds up the process and allows the use of standard paragraphs.

Some software systems allow the surveyor to dictate specific action points within the report and then collect these together at the end and automatically populate a list in the form of a summary with a reference to the relevant sections in the report. So at the conclusion of the report the client has the basis of a specification of works. Some surveyors take this a step further and include a list of approximate building costs.

I do not recommend that approximate building costs be included in a Building Survey report. There are two reasons. First in order to price the work accurately the surveyor will need to measure the work involved and prepare approximate quantities which will involve time and increase the fee. It may be that upon receiving the report the client will decide not to proceed with the purchase, in which case the need for estimates will not arise.

Secondly, there are many items of building work which can be undertaken in a variety of ways depending upon the client's requirements and in the absence of full knowledge of the client's needs it may be impossible to assess costs accurately. If estimates are provided the Royal Institution of Chartered Surveyors (RICS) Guidance Note recommends that the surveyor should be careful to state at some length the reservations and limitations of such advice.

It is obvious that if a surveyor advises a client that a particular defect will cost £10 000 to remedy and the actual cost turns out to be £20 000 the surveyor is vulnerable to a claim for negligence. It is a matter of common experience that the cost of building works invariably turns out to be more than the budget especially when dealing with old buildings in poor repair.

Disclaimer and exclusion clauses

It is normal for a Building Survey report to include disclaimers of liability or exclusion clauses and the professional indemnity insurers will probably insist on this as a condition of providing insurance cover.

The distinction between a disclaimer and an exclusion might best be seen in relation to a dry cleaning shop: if there is a notice on the wall saying 'We do not guarantee to get your clothes clean' this would be an attempt to disclaim liability for poor results whereas if it says 'We do not clean the right sleeves' this would be an attempt to exclude liability altogether for that part of the garment.

All such disclaimers and exclusion clauses are subject to the test of reasonableness under the provisions of the Unfair Contract Terms Act (see Chapter 19).

The following are typical examples of standard clauses which may be inserted into report to limit liability and provide standardised advice:

- *Limits of inspection*: At the time of inspection the weather was warm and dry and the house was occupied and furnished with floors covered. Fitted floor coverings have been lifted in sample areas only in order to identify the nature of the construction beneath and we have not removed fitted floor coverings generally. Similarly loft insulation material has been disturbed only to the extent necessary to confirm the nature of the construction beneath and we have not removed loft insulation material generally. We have not inspected inside walls cavities or seen the wall ties. No exposure of hidden parts of the structure has been undertaken.

- We have not inspected any parts of the structure which are covered, unexposed or inaccessible and we are therefore unable to report that such parts are free from rot, beetle or other defects.
- We inspected inside the subject flat and the common parts giving access to it together with the external elevations from ground level only and we have not seen inside any other flats in this building or any roof spaces.
- *Foundations*: We have not undertaken any trial bores in order to confirm the nature of the subsoil under this property, however, the Geological Survey Map for the area indicates that the subsoil is likely to be (xxxxxxxxxxxx).
- We have not dug any inspection pits in order to examine the foundations, however, houses of this type and age in this locality were usually constructed using conventional shallow strip foundations consisting of a concrete strip with brick footings typically laid about 600 mm to 900 mm below ground level. This is a fairly shallow foundation by modern standards.
- When buildings have shallow foundations on shrinkable clay subsoils it is generally advised that no trees or bushes should be planted closer to the main walls than their mature height because tree roots, extracting moisture from the clay subsoil, can cause damaging foundation movements.
- *Cements and concretes*: No tests have been undertaken on cements and concretes used in the construction and we are therefore accordingly unable to confirm the presence or otherwise of high alumina cement, chlorides, sulphates or other deleterious materials. As regards concretes below ground we cannot confirm that they will be suitable for the ground conditions if the subsoil contains sulphates or other damaging constituents.
- *Roofs*: Our inspection of the roofs externally has been from ground level, or using a surveyor's ladder where roofs are no more than 3 m above ground. Some parts of the roof cannot be seen due to the angle of view available.
- *Flues*: If you wish to use flues for open fires or appliances they must first be swept, the condition of the linings confirmed and a smoke test carried out. In an old building constructed in a soft lime mortar it is likely that flues would need to be relined before they can be safely used and a reputable fireplace contractor should be consulted. If flues are redundant they should be provided with air-vents at the base and hooded pots at the top to keep them dry.
- *Gutters*: Periodic checks of gutters and downpipes are advised during heavy rainfall to ensure that roof water is being taken away freely without leakage. This house has solid walls (in contrast to modern cavity construction) so any leakage of water down the walls externally can penetrate directly to the inside.
- *Timber defects*: Woodworm and other timber defects are commonly found in houses of this age in this locality. Local timber treatments and repairs will be required from time to time. This should be regarded as normal ongoing maintenance. Timbers are less likely to suffer such defects if kept dry in a centrally heated house and if the sub-floor ventilation is unobstructed.
- *Condensation*: Condensation is a common problem in this type of property. It is important to maintain the correct balance of background heating and ventilation. The temptation to dry laundry, etc. on the radiators should be avoided and some form of extraction ventilation should be provided to bathrooms and kitchens notwithstanding that these rooms may have windows. So an extractor fan, operating automatically from the light switch with a time delay is advised in the bathroom and a ducted cooker hood would be much to be preferred as compared to a simple filter hood in the kitchen.

- *Double glazing*: A problem often arises due to condensation forming between the glass in sealed units and when this happens the only remedy is to reglaze the windows which can be expensive. You should ask your legal adviser to confirm when the double glazing was installed, who did it, and whether it is covered by a long-term transferable guarantee.
- *Electricity*: An old consumer unit with fuses is located under the stairs. We recommend that the system be inspected and tested by a National Inspection Council for Electrical Installation Contractors (NICEIC) registered electrician tests to include insulation, polarity and earth continuity with a check to ensure that all plumbing and gas services are bonded to earth. Any works necessary to comply with British Standard 7671 should be undertaken. We advise replacing the fuses with a modern residual current circuit breaker (RCCB) unit and circuit breakers.
- *Gas*: We recommend that the gas service together with any gas appliances included in the sale be inspected and tested by a Council of Registered Gas Installers (CORGI) registered engineer paying particular attention to boiler flue and ventilation requirements.
- *Water*: We have not seen the main underground supply pipe from the pavement but if this is an old lead pipe it will now be due for renewal in polypropylene and this will involve digging up the front garden. Further enquiries with the vendor are advised to ascertain what is known of the age and type of main water supply pipe.
- *Central heating*: We recommend that the central heating system be inspected and tested by a CORGI registered engineer and a report on its condition obtained. Advice should be sought to confirm whether the system needs to be chemically cleansed and recharged with an inhibitor.
- *Drainage*: We inspected inside manholes within the curtilage of the subject property only and no tests have been undertaken. The drain flow within the manholes was satisfactory at the time of inspection.
- *Tenure*: The advice given in this report has been prepared on the assumption that the property is freehold with an unencumbered absolute title and with vacant possession. Some of the property titles in this area incorporate restrictive covenants which require an original freeholder's consent for alterations or extensions and you should ask your legal adviser to confirm whether any restrictive covenants apply in this case. You should also ask your legal adviser to confirm who is liable for the various boundaries by reference to the title plan and covenants.
- *Converted flats*: The standard of sound-proofing between flats is likely to be poor and you may rely for your quiet enjoyment of the flat on the consideration and cooperation shown by the neighbours above and below. Complains regarding noise nuisance are commonly made by the occupants of converted flats with timber floors. You should ask your legal adviser to confirm that there are suitable lease covenants against noise nuisance such as the playing of musical instruments at unsocial hours, etc. There should also be covenants that floors have to be kept carpeted. The modern trend to have exposed floor boards or laminated wood flooring in flats can result in considerable noise nuisance to the occupants below.

The above are just a few examples of simple standard text; in most surveyors' offices the amount of standard text is far more extensive covering the whole range of building types and defects encountered. The important thing to remember with standard text is that it should be altered and adjusted in each case to suit the specific property and it should read well so that the report flows and the client is not left with the impression that the whole thing has been put together by a computer.

Subsequent dealings with clients

It is quite common for clients to wish to discuss the report. Clients often use the report in negotiations with the vendor in the hope of agreeing a price reduction to reflect the defects found. The surveyor may be drawn into further correspondence regarding the property. It is important to ensure that any letters, e-mails or other correspondence emphasise that the original terms and conditions applicable to the report continue to apply and care should be taken not to make off-the-cuff remarks intended to be reassuring to the client which could rebound later. This would apply, especially to any discussion of building or repair costs or the ease with which particular defects could be remedied.

Complaints

All Chartered Surveyors are required to have a complaints handling procedure in place and in most cases it is prudent to arrange for another surveyor to visit the property to investigate a complaint, rather than allow the original surveyor to go back. In my experience most complaints are not about defects in the property but about people and their varied reactions to circumstances. Often the complaint can be resolved by explaining to the clients precisely what was included, and what was not included, in the terms and conditions of engagement which they signed.

11
Home condition, homebuyer and other pro forma reports

Terminate or postpone the visit if the seller/occupier attempts to intimidate or abuse you in any way. Do not allow the occupier to lock doors behind you. Do not inspect after dark or at dusk. Do not enter rooms where persons are sleeping, scantily clad or obviously drugged or drunk. If the home is only occupied by children or juveniles, you must arrange to postpone your visit until a responsible adult is present.

Section 2.5.3. Home Inspector's Inspection and Reporting Requirements, Version 3, February 2005

A distant relative of mine was a surveyor in Denmark before and during the Second World War. He went around the Copenhagen area preparing mortgage valuations for what was then a Danish equivalent of one of our Building Societies. No doubt the reports were very similar to those prepared in the UK at the same time although the construction was rather different and a lot more timber was used in Danish house construction.

For the last 12 years the Danes have incorporated into their house selling process the concept that the seller has to prepare, or have prepared, a seller's pack of information including title details and other matters which the purchaser's legal representative is then given to speed up the sale. In theory, if there is no chain and it is a cash purchase, the buyer could view the house in the morning and buy it in the afternoon.

Encouraged by the Danish experience it has been a long-standing New Labour promise to introduce Sellers Packs in the UK and at the time of writing the 6th edition of this book the necessary legislation is in place and the Information Technology (IT) and training are well advanced.

A particular feature of the new Home Information Packs (HIPs) is that they will include a Home Condition Report (HCR) which will itemise the various parts of the property and give each part a Condition Rating.

Home Condition Reports

The Housing Act 2004, Regulations made under the provisions of that Act and the National Occupational Standards for Home Inspectors taken together provide a framework for the provision of HCRs. HCRs will form part of the HIPs which it is intended will be prepared in respect of most residential properties being placed on the market for sale in England and Wales. At the time of writing the proposed date for the introduction of compulsory HIPs in England and Wales will be 1st June 2007.

The commentary which follows is based upon the information available at the time of publication of this book and may be subject to changes as the scheme evolves. There is still a considerable amount of political controversy in relation to these proposals.

As currently proposed the scheme does not apply to non-residential property – factories, offices, shops warehouses, etc. – nor will it apply to mixed commercial and residential properties so a shop with living accommodation above will not be covered. Leaseholds for 21 years or less and residential investment properties subject to tenancies will be excluded.

Experienced surveyors will find that the way in which an HCR is put together will be quite different from the familiar Building Survey or Homebuyer Report. And no matter how experienced the surveyor it will still be necessary to undertake the training and assessment in order to qualify as a Home Inspector. The two important new distinguishing features of the HCR compared with a Homebuyer, for example, will be that the report is heavily focussed on providing a Condition Rating for the various elements inspected and that the Home Inspector will be acting for the vendor but with a potential liability to the vendor, the purchaser and also any mortgage lender who relies upon the report.

So when a house or flat is being placed on the market for sale it will be a pre-condition of marketing the property that it is inspected by a Home Inspector and an HCR is prepared. The Home Inspector will be instructed either directly by the vendor or by an estate agent acting for the vendor and they will be employing an HIP Provider to put together the legal and other documentation of which the HCR will form a part. The Home Inspector will need to make arrangements with the owner of the property to undertake an inspection and complete a report that meets the necessary reporting requirements.

Within the HCR the various parts of the property are listed and one of four Condition Ratings are applied:

Condition Rating	Definition
Not inspected	Self explanatory
1	No repair is presently required. Normal maintenance must be undertaken.
2	Repairs are required but the Home Inspector does not consider these to be either serious or urgent.
3	Defects exist of a serious nature or defects requiring urgent repair.

A present or potential defect that requires further investigation is reported as Condition Rating 3. A serious defect is one which is likely to compromise the structural integrity of the property and/or affect the health and/or safety of the occupiers.

A defect requiring urgent repair is typically one likely to develop rapidly into a serious defect if not repaired/remedied immediately and/or cause structural failure or serious defects in other building elements if not repaired/remedied immediately.

The report includes an energy rating which satisfies the requirements of the European Union (EU) Directive 2002/91/EC of 16th December 2002 using the Reduced Data Standard Assessment Procedure (RDSAP) so in addition to the inspection for defects some measurements will need to be taken and a pro forma RDSAP form completed. The report also includes an estimated re-instatement cost for insurance purposes using floor areas calculated in accordance with the Royal Institution of Chartered Surveyors (RICS) Code of Measuring Practice (gross external area for houses, gross internal area for flats).

HCRs do not have a time limit. Caveat Emptor (buyer beware) continues to apply in the home-buying process and it is, therefore, up to the purchasers to decide whether they are happy to rely upon an HCR which was prepared some time ago, especially if the property has been on the market for a long time.

When a property has been taken off the market for more than 28 days and is then placed back on the market a fresh HIP with a new HCR has to be prepared. So a vendor who is uncertain whether to take the house off the market or not would be advised to keep it on the market, at least nominally, until he or she is sure otherwise the additional cost of a fresh HIP may arise.

Home Inspectors

The requirement for Home Inspection Reports (HIRs) has resulted in the need to train Home Inspectors. Home Inspectors will generally qualify to VRQ or NVQ level 4 via an ABBE Diploma in Home Inspection. The training will be focussed on HIRs only and will not include Building Surveys of the type described in this book, or other forms of pro forma report such as the RICS Homebuyer.

Home Inspectors will need to have professional indemnity insurance and they will be potentially be liable to the vendor, purchaser and a mortgage lender in respect of the contents of their reports. The inspection will require the completion of pro forma site notes and it will be necessary for Home Inspectors to keep comprehensive records of inspections. Lifetime learning in the form of Continuing Professional Development (CPD) will be required in addition to the initial training so the Home Inspector will need to maintain a satisfactory record of CPD.

The Home Inspector will be employed by, and ultimately paid by, the vendor (even if indirectly through an HIP provider). But he or she will visit the property with the intention of finding faults so the whole approach to the inspection and the preparation of the report will require a degree of sensitivity and diplomacy especially if the property is in poor condition with extensive defects and evidence of neglect. In this respect the task becomes more difficult than would be the case with a Building Survey or Homebuyer Report where the surveyor is acting for the buyer alone and the vendor normally accepts that the surveyor will not discuss his or her findings or the contents of the report which is confidential to the buyer until such time as the buyer wishes to disclose it.

Home Inspectors will be expected to identify construction types which are unusual and with which they may be unfamiliar. In such cases they are required not to complete the HIR but refer the case to another Home Inspector with the necessary experience. This is likely to be an issue with some system-built houses and flats (mostly, but not exclusively ex-local authority) or old timber frame or cob construction.

The particular skill which the Home Inspector will have to acquire will be the ability to allocate the appropriate Condition Rating. Generally Condition Rating 1 will be fairly straightforward and a Home Inspector operating in an area of mostly modern estate developments may well find that reports will routinely issue Condition Rating 1 to all items.

Equally those serious defects, or items requiring further investigation, which need a Condition Rating 3 will be fairly obvious. An old slate roof without underfelting largely held together by gravity and tingles, with pigeons nesting in the loft, is clearly destined to be Condition 3.

The skill of the Home Inspector will be to allocate the Condition Rating 2 because here we are into a grey area of judgement and opinion. A distinction has to be made between normal maintenance (Condition 1) and when repairs are required but they are not serious or urgent (Condition 2). Equally looking from the other direction there may be defects which the Home Inspector has to judge either serious (Condition 3) or not serious (Condition 2).

The scheme allows for the seller to appeal against the findings of the Home Inspector and request that another Inspector be appointed to prepare a fresh report if he or she is unhappy with the decisions made.

The Home Inspector should not make any comment on matters of taste or decoration or on the suitability of the location of the property in relation to local amenities. No opinion on the value should be included. The HCR should be strictly limited to the parameters of the report format and the construction and condition of the property. It is up to the purchasers to satisfy themselves regarding the location in relation to schools, transport and other facilities and to allow for improvements to decorations or fittings to suit their own tastes.

Fig. 11.1 The Home Inspector should not comment in the HCR on questions of taste.

Homebuyer Reports

In 1981, the RICS introduced the Homebuyer Report (then called the House Buyers Report and Valuation) which was described at the time, and sometimes since, as a sort of mid-way report between the simple Mortgage Valuation Report on the one hand and the full Building Survey on the other. A number of other rather similar pro forma reports have since appeared including the Home View Report sponsored by Abbey Plc.

The standard of inspection required for a Homebuyer Report will generally be much the same, however, as for a Building Survey as described in this book but subject to certain specific limitations laid down in the conditions of engagement. The main distinctions are that the Homebuyer Report should be very brief, using a pro forma layout, and that generally only significant defects or urgent repairs are reported.

In *Cross and Another* v. *David Martin and Mortimer* (1989) IELGR 154 Mr Justice Phillips, sitting in the Queens Bench Division of the High Court, had to consider a case concerning a Homebuyer Report.

The plaintiffs had alleged that the surveyors were negligent in that they had failed to draw attention in the report to three features of the property; settlement of solid ground floor slabs, misalignment of a number of first floor door openings and alterations which had weakened roof trusses when the loft was converted into a room. The surveyors did not dispute the existence of these defects but contended that the first feature was not ascertainable during a Homebuyer inspection (which excludes lifting fitted carpet) and the other two features were not sufficiently serious to warrant comment in the report.

The surveyors lost the case. It is unwise to read too much into the result of one case based upon its own set of facts, nevertheless the outcome would appear to confirm that the Courts would view a Homebuyer report as a Building Survey presented in an abbreviated form rather than a simple valuation with limited added comment on condition.

So whilst the report itself may be brief and defects which are not urgent or significant may go unreported, nevertheless the Homebuyer, in the scale of things, is not really a mid-way report but rather closer to the Building Survey than it is to the simple mortgage valuation.

One of the possible side effects of the introduction of HCRs will be a considerable reduction in the demand for Homebuyer Reports prepared for purchasers since it is expected that in many cases the purchasers, having been presented with an HIP which includes a generally favourable HCR will be comfortable with that and will not require their own report. On the other hand, there could be an increase in the demand for full Building Surveys from purchasers who have an HCR with a large number of Condition Rating 3 needing more information regarding the scale and extent of the defects.

Mortgage Valuation Reports

Some years ago there was an attempt by the RICS and the Council of Mortgage Lenders to agree a standardised form for Mortgage Valuation Reports in the hope that the same form might be used by all the mainstream banks and building societies. What bliss it would have been for mortgage valuers and the support staff in their offices if such a form had been agreed! Unfortunately too many cooks spoiled this particular broth and the best efforts of those concerned failed to produce a result. As a consequence there are currently hundreds of forms used by the various mortgage lenders in the UK and valuers who regularly report for a large number of clients are faced with a multiplicity of reporting requirements.

All this may be about to change, however, because the Council of Mortgage Lenders has resurrected the idea in the form of a simple one page valuation report which could be prepared in conjunction with the HCR. There is also likely to be a considerable reduction in mortgage valuations generally in future years because many mainstream mortgage lenders are now moving towards using software systems to verify valuations, resorting to 'desk top' valuations or relying upon very brief external inspections of mortgage securities.

On the following pages I reproduce a copy of one of the better designed forms which is actually based upon the original RICS proposals and used by the Nationwide Building Society. The way in which the form needs to be completed is mostly obvious but there are a few issues arising which are worthy of comment.

In keeping with the usual procedures of the UK Building Societies and Banks the form is issued in triplicate with one copy for the lender (mortgagee) one for the borrower (mortgagor) and one for the valuer's file. It will be noted that there are important notes for the borrowers in the terms and conditions they sign being exclusion clauses intended to limit the liability of the valuer to the borrower. There should be no doubt in the mind of the borrowers as to the very limited nature of the inspection and the purpose of the report. Nevertheless such clauses are subject to the test of reasonableness under the provisions of the Unfair Contract Terms Act and the judgment of Mr Justice Park in *Yianni* v. *Edwin Evans and Sons* (see Chapter 19).

It will be seen that the valuer has to declare that he or she is not contravening Section 13 of the 1986 Building Societies Act which is concerned with potential conflicts of interest. So the valuer must have no interest directly or indirectly with the property being offered as security.

The RICS publishes Statements of Asset Valuation and Guidance Notes (The Red Book) the provisions of which are mandatory for Chartered Surveyors carrying out this type of work.

'Desk Top' and 'Drive Past' Valuation Reports

Since April 2000, the Land Registry records of all property transactions in England and Wales have been available and there are now a number of web sites where you can find out what your neighbours sold their house for. In addition to being a boon for the nosey and curious that has been a very welcome development for mortgage valuers who can now access lots of transaction details by address or postcode from a huge and growing database.

Mortgage lenders have come to the decision that if the property is of fairly standard type in an area where there is lots of sales evidence and if the loan-to-value ratio is reasonable, they do not need a full

| Report Number | | Instructing Office: | |
| Voucher Number | | | Reporting Office: | |

Residential Property Mortgage Valuation Report for Nationwide Building Society

Inspection date: _____

1. Mortagage Details

Applicant(s) Name(s) _____

Property Address inc. postcode _____

Postcode _____

Purchase Price £ _____   Advance Amount £ _____   Term _____ Years

2. Tenure

Tenure   Freehold (0) Leasehold (1) or Feudal (2)   If Leasehold unexpired term _____ Years

Rent – Ground/Chief/Feu £ _____ p.a.   Fixed/escalating etc. £ _____

Maintenance charge £ _____ p.a.   Other _____

If shared ownership, what % of the property is being purchased   _____ %

3. Tenancies

Is there any tenancy apparent?   Yes/No   If yes, please give details and rent(s) upon any agreement _____

4. Property Description

Type of Property   Detached House (01) ☐   Semi-Detached (02) ☐   Terraced House (03) ☐   Detached Bungalow (11) ☐

Other Bungalow (12) ☐   Purpose Built Flat (31) ☐   Converted Flat (33) ☐

Approx. Year Built _____   If a flat, which floor is it on? _____   Is there a lift?   Yes/No

Number of Bedrooms _____   Number of Bathrooms (inc. ensuite bath/shower rooms) _____   Number of habitable rooms _____

5. Property Construction

Main Building Construction – Walls _____   Roof _____

Garage/Parking Space   Single (1) / Double (2) / Parking Space (3) / None (4)

Is the parking space/garage offsite?   Yes/No

6. Services

Mains Services Available   Electricity / Gas / Water / Drainage / Other

Type of Central Heating   None (0) / Full Gas (1) / Full Electric (2) / Full Oil Fired (3) / Full Solid Fuel (4)

Part gas (5) / Part Electric (6) / Part Oil Fired (7) / Part Solid Fuel (8)

7. New Properties (if applicable)

Name of Builder _____   NHBC / Zurich / Architect / Other

Are roads / footpaths made / partly made / unmade   Estimated cost of making up £ _____

8. Buildings Insurance

Estimated current re-instatement cost including site clearance and professional fees excluding VAT. except on fees £ _____

Floor area – Main Building _____ m^2
Garage if not integral _____ m^2
Other Buildings _____ m^2

Does this property need to be referred due to special insurance risks?   Yes/No   If Yes include appropriate key statements in Section 10 – General Remarks

9. **Other Matters that may Materially Affect value**
    (If applicable give more detail in the general remarks below)

Is the property readily resaleable at or about the valuation figure?	Yes/No	In the case of flats etc. is proper management / maintenance apparent?	Yes/No
Has the property ever been affected by structural movement caused by subsidence, settlement, landslip or heave	Yes/No	Is the risk of further movement on the Society can accept? (If No decline property)	Yes/No
Rights of way / Easements / Servitudes / Wayleaves (where apparent on inspection)	Yes/No	Building works that may have required Planning Permission / Building Regulation approval	Yes/No

Any other important factors?   Yes/No   If Yes give details:

10. **GENERAL REMARKS** (including the general condition of the property)

11. **WORKS TO BE CARRIED OUT** as condition of mortgage subject to retention below. (Listing should only include work absolutely necessary to protect the society's security. The amount of advance must be ignored)

Amount of Recommended Retention (Minimum retention amount is £1000. This is not an estimate of costs. The Applicant(s) should obtain detailed estimates before proceeding with the purchase)   £

12. **VALUATION FOR MORTGAGE PURPOSES** – (assuming vacant possession unless otherwise stated)

Is the property a suitable security for the Society   Yes/No

If Yes, valuation in present condition   £

Valuation upon completion of any works required under section 7 or 11   £

If shared ownership, value of share being purchased   £

I certify that I have personally inspected the property and that in making this report I am not contravening Section 13 of the Building Societies Act 1986, or any variation or re-enactment of it.

Valuer's Signature

Name and Qualification
Firm's Identity Code

Date

Firm
Address

Postcode
Telephone
Fax Number

IMPORTANT NOTICE TO APPLICANT(S)

This report has been prepared solely for the Society's purposes. It is not a structural report and is based upon a limited inspection. It may not reveal serious defects and may contain inaccuracies and omissions. It is unlikely to be adequate for a purchaser's purposes and should not be relied upon.

YOU ARE STRONGLY ADVISED TO OBTAIN A FULLER REPORT ON THE PROPERTY.

The Society does not guarantee that the purchase price is reasonable.

internal inspection and a detailed report. So the Nationwide form I have reproduced, and other similar forms, may in time become largely redundant and used only in those cases where the property is unusual or the borrowing very high in relation to the value.

So the property could be valued using a computer programme of which there are a number now, growing in sophistication all the time. Or the valuer could be asked to complete a simple valuation report based either upon a desktop assessment or a brief external inspection of the property. The terms and conditions of engagement for such reports set out clearly the limitations of the valuer's liability.

One of the reasons mortgage lenders have moved away from the more detailed inspections and reports has come about as a consequence of experience during the recession in the 1990s when many properties were taken into possession and they lost a lot of money. The fact that a valuer had inspected the house and prepared a fairly detailed report on it did not, in the event, make any difference to the loss. Overwhelmingly the loss arose because the borrowers could not afford to pay their mortage and the property market dropped. So whilst it would obviously be prudent to have some check on the price or valuation – if for no other reason than to pick up frauds – a report with lots of information about the property is largely unnecessary.

Some of the information provided in these reports is used by the mortgage lenders for statistical purposes. For example, both the Nationwide and the Halifax prepare a house price index based upon the data collected from these forms.

Energy Rating Reports

Surveyors will become increasingly familiar with the need to complete Energy Rating Reports on residential property as part of the Government's drive to improve the energy efficiency of our homes and reduce CO_2 emissions. A RDSAP report will automatically form part of the HCR and is already required along with the valuation by some mortgage lenders.

The surveyor completes a simple pro forma form on site which is then downloaded to an energy rating company who produce a report. Essentially these reports are in two parts; first the actual heat losses and CO_2 emissions from the house or flat are calculated using the software system, then the report will include some advice as to how heat losses and CO_2 emissions can be reduced. Normally the recommendations made are listed in order of cost/benefit.

Loft insulation usually comes first with up to 250 mm of fibreglass quilt or similar currently advised for optimum benefit. So if anything less than this is reported an upgrading of loft insulation is usually advised. Then we have draught proofing – lots of benefit for minimal cost. Then perhaps double glazing, solid wall lining, cavity wall insulation, floor insulation and other improvements are suggested.

The effectiveness of the heating system is also included in the analysis so if there is an old, relatively inefficient, boiler an upgrade to a modern condensing boiler may be advised. Surveyors who obtain RDSAP reports on a regular basis will know that the standardised advice will not always be suitable in every case and the implementation may be complicated by planning regulations for listed buildings or in conservation areas. In the case of flats the management arrangements may preclude individual owners changing windows or carrying out insulation work.

Properties in possession

Brief pro forma reports are prepared for mortgage lenders on properties in possession usually using standard forms. These are commonly called 'repossessions' although strictly speaking since the property has never been in the possession of the mortgage lender the terminology is incorrect. Whereas with most Mortgage Valuation Reports the liability of the valuer for negligence arises if the valuation is too high, with possessions the opposite is true and the liability to both lender and borrower arises if the valuation is too low.

Fig. 11.2 The Trellick Tower in west London, now a listed building. The preparation of a brief pro forma report on a flat here presents some challenges.

The surveyor in these circumstances may be required to provide a range of opinions as to the value assuming short and long periods of marketing, and to advise regarding the best method of sale which may be private treaty, auction or tender. It is worth remembering that the defaulting borrower has rights and if the eventual sale price exceeds the debt and costs the borrower may be entitled to some of the proceeds of sale so it is most essential that the surveyor provides sound marketing advice.

More often than not, however, there is a shortfall on sale so the mortgagees in possession may be chasing the defaulting borrower for their loss. In these circumstances the borrower (or more likely his or her lawyer with the benefit of legal aid) may challenge the sale price and allege that the property was sold at an undervalue. For this reason it is all the more essential that the surveyor keeps good records of the inspection, the comparable evidence used and all other pertinent information.

Part Exchange Reports

These pro forma reports are generally prepared for builders selling properties on new estates, or for chain-breaking companies and here again the surveyor is usually providing some marketing advice in addition to a valuation. Watch out for non-standard construction or unusual properties for which there may be a limited market. Human nature being what it is vendors will often hope to unload a potentially difficult property by way of a part exchange deal if they are finding it difficult to sell in the open market in the normal way.

12

Reports on non-residential buildings

We're taking a full repairing and insuring lease. They are in rather a hurry so I have agreed to sign the contract this afternoon. I thought that you could just give it the once-over and 'phone me back before lunch. Incidentally, he says that the flat roof has been leaking but it's all right now.

Potential client's telephone instruction

The basic principles described for surveying and reporting upon dwelling-houses also apply to other types of building, the main distinctions being those of size and usage, although a larger industrial building may, if modern with open headroom and clear spans, be easier to inspect than a small house.

The larger older buildings will often require a lengthy inspection, perhaps lasting several days and the surveyor will need to ensure before undertaking such work that he or she can reserve sufficient time in the diary to complete the inspection to the necessary standard.

A special consideration arising in the case of the larger commercial buildings is the importance of concealed structural elements. Before a client can be advised on the construction of such a building and before any structural alterations can be contemplated, it is essential to determine the structural strength of the building, including the manner in which the loads are carried and transferred to the subsoil.

Where concealed structural elements are involved, the surveyor will have to undertake an initial inspection and may then have to proceed with a process of research and detection in order to discover the facts. Such research will normally include searching for any existing architects' or engineers' drawings (in one case a most useful set of such drawings were found in a wall cavity), visiting the local authority offices or district surveyor to see if any drawings are available for inspection, and even perhaps opening up the building in question to confirm the position.

If the building has an engraved plaque or stone with the name of the original architect indicated, then the first step would be to see if he or his firm can be traced. The present occupiers or owners of the building may have been there since completion in which event they should know the identity of the original designer and perhaps even own a set of plans. Even if not the original owners, it is possible that plans, or information on the location of plans, may have been passed on to them by their predecessors.

If the present owners are a large company, contact with a long-serving member of the staff may help since such an individual may provide useful information if he can recall the building during construction. For example, the names of builders, architects, manufacturers of fittings, sub-contractors and other vital information.

Working drawings for steelwork may be obtained if the identity of the original steelwork sub-contractor can be discovered, since such contractors often keep records going back for many years. A large number of steel-framed buildings were erected between 1925 and 1949 where such a sub-contractor, if located, may be the only likely source of information.

Having exhausted these sources of original drawings, the surveyor could then go on to see if information is available from the local authority. Local authorities are now much more reluctant than in the past to permit an examination of their records. Such information as is available is often microfilmed and during that process, many older plans and documents may have been destroyed.

From the local authority point of view, there are two problems affecting a general inspection of their files and records.

First, it is considered that in order to avoid a breach of copyright no copying of deposited plans of any kind should be permitted without the original owner's or architect's consent. In order to take sketches from a deposited plan one has, on occasion, been required first to obtain a signed authority from the original author of the drawing. In many cases obtaining any such authority is impossible, particularly in the case of older buildings.

Secondly, local authorities are becoming increasingly subject to litigation over building defects and will often fear that attempts to examine closely approved plans may result in legal actions should any errors or omissions be found. A general reluctance to permit any inspections at all has, as a consequence, arisen with local authorities in some areas.

A surveyor who is a member of the local authority staff will, therefore, be in an advantageous position in this respect when investigating and reporting for his or her authority upon buildings in his or her area.

Having thus obtained all available information in the way of plans and working drawings and having carried out an initial survey of the building concerned, the surveyor will then have to consider the extent to which the building ought to be opened up in order to complete the report.

By this stage, it will have been possible for the surveyor to have formed a view of the superficial condition of the building and services, also the extent to which it complies with current standards for means of escape in case of fire, fire precautions, the Factories Acts, the Offices, Shops and Railway Premises Act and other legislation so that it may be that a preliminary report could be prepared for the client on this basis. If so, it is essential that the client be warned that no opening up of the structure has been undertaken and that the condition of the main structural elements has not been confirmed.

Post-war office blocks and other structures constructed in reinforced concrete present a particular problem and considerable care needs to be exercised if the structure has to be disturbed. Reinforced columns and beams cannot be opened up easily or exposed to see how reinforcement is arranged and such reinforcement is essential to the strength of the structure. If pre-stressed beams are suspected, it is most necessary that these should not be cut into or otherwise disturbed. In such circumstances, a surveyor should measure the overall sizes and spacings of the columns, the dimensions and spans of the beams and then undertake a surface inspection for evidence of chemical deterioration and like problems. Beyond this the surveyor would be bound to refer his or her client to a specialist consulting engineer so that any core samples may be taken or special tests made to detect the presence of high-alumina cement (HAC), chlorides, sulphate attack or other such faults having regard to the known facts.

Steel-framed office buildings and other structures dating from the inter-war and post-war years may be opened up more easily than reinforced concrete structures, and the quality of the original construction and the condition of the steelwork may then be judged. It is arguable as to how far one should take such investigations but if the surveyor can confirm sizes, spacings and spans of principle members, together with the loads imposed, it should be possible for him to obtain a check by calculation from a structural engineer and thus reassure his client that the steel-framing is technically adequate and, where seen, in satisfactory condition and protected against corrosion.

It is clearly essential for the client to be advised whether or not a commercial or industrial building will satisfy the Fire Precautions Act 1971 and the Fire Precautions Act 1971 (Modifications) Regulations 1976, together with any subsequent or additional legislation having regard not only to the results of the building inspection but also to the specific needs of the particular client and the use he proposes to make of that building. The provision and maintenance of fire precautions and means of escape form an increasingly important part in the day-to-day management of commercial and industrial buildings. If a need for certification under the Act arises, whether from a proposed change of use, structural alteration or other reason, the client will need to know in advance whether or not difficulties in obtaining certification could arise and exceptional costs occur in adapting the building to meet certification standards.

It is necessary to consider whether or not exceptional energy costs are involved in lighting and heating the structure under review, and the surveyor should then inspect the existing space heating appliances. Generally, a specialist report from a heating engineer should be recommended and some investigation then carried out into past energy costs. In most cases, the vendor or existing lessee of the building will be able to provide details of accounts for oil, gas, electricity and other energy inputs into the building and these may be used as guidance to current cost levels as the building is presently used. Care should be exercised when assuming that current energy cost levels will apply to that building in the future, and if the client's needs change then a change in energy costs may also arise.

At this point the surveyor may have to advise the client on the overall condition of energy-intensive services and whether or not they are efficient, or if the type of energy in use might be changed for a more economic type (e.g. gas space heating instead of oil, assuming this can be made available). One should not assume that current relative differentials in energy costs will always be maintained. Energy which is proportionately cheap now need not always remain so and the capital cost of a change-over may not be economic if the differential in cost becomes less advantageous in the future.

As far as energy costs are concerned, the overall thermal efficiency of a building is most significant and should be considered by the surveyor. An industrial structure with a high eaves level, ideal for loading bays and overhead gantries, will be very wasteful of heat as compared with a similar floor space with low headroom. If there is no need for a high ceiling, then the provision of a suspended ceiling could be recommended to reduce space heating costs. This is particularly applicable to a single-storey structure or the top floor of a multi-storey structure when, in addition, insulation over the suspended ceiling may be suggested.

When considering the services, thought should be given to the client's proposed use of the building and the numbers and sexes of staff likely to be employed therein. Recommendations may then need to be made over the suitable provision of lavatory and washing facilities. The Offices, Shops and Railway Premises Act 1963 demands sufficient and suitable lighting, a reasonable temperature in every room (15°C), sufficient washing facilities, a supply of hot and cold water, and adequate ventilation. It also covers a variety of other environmental needs. Insofar as the premises inspected may fall short of the standards required, the necessary comments and recommendations should then be made.

Older commercial and industrial buildings are generally found to have been constructed in a great variety of ways. There is often a mixture of cast-iron, wrought-iron or steel columns and beams forming the basic structure, and load-bearing brick or stone walls to the exterior and around stairwells and shafts, with timber or concrete floors of various types spanning between the main elements. The concrete floors may be constructed using obsolete methods, often in the form of *in-situ* poured concrete construction or part pre-cast and part *in-situ* construction. Research into the various forms of construction which are found is a most interesting area of professional activity, particularly in view of the great variety of construction methods likely to be found.

Floors in older commercial and industrial buildings are often of filler-joist construction with a layout of steel or iron beams and an infilling of *in-situ* concrete, the beams being either of solid or open-webbed types. In those structures erected before the advent of modern Portland cement concretes, it is likely that a lime mortar could have been used to bind together a clinker, brick or stone aggregate. Some caution is required in assessing the strength of these older types of lime mortar solid slabs since they will be most unlikely to have a strength comparable to the modern Portland cement concrete slabs. This may be significant if an older structure is to be used for different or heavier machinery.

Opening up older structures may be necessary to determine the sizes of the beams and columns, the thicknesses and strengths of load-bearing walls and the nature of the floor construction, including the effective spans of the flexural members and their manner of bearing. Finally, having traced the main forces and loads through the structure down to ground level, a final confirmation of the size, type and adequacy of foundations used may be needed having regard to both present loadings and any different future loadings which may apply. Samples would then be taken of concrete, bricks, mortar, steel, wrought iron, cast iron and timber to confirm the condition of such elements, and these can be sent to specialist laboratories for analysis with a view to confirming the strength of the structure.

In addition to confirming the construction of the building and the nature of the materials used, it is also necessary to confirm the condition of framing since this will be relevant to the strength. Obviously, if an engineer is to be asked to check the suitability of a building by calculation on the basis of the information given as to the loads, spans and materials used, any such calculations will be valid only if the main structural elements are in good condition. Samples taken may or may not be representative of the structure as a whole.

In many old buildings, ironwork or steelwork is buried in the walls in a manner which provides only limited protection from the elements and corrosion may be severe, especially on the weather sides or at points where leaking gutters or other sources of dampness have subjected the walls to constant saturation over many years. The older concretes, consisting of breeze and lime mortars, contain a high sulphate content which will attack adjoining ferrous metals. Roof leaks may cause deterioration to steel roof joists and the upper parts of a steel- or iron-framed structure. Steel columns set into basement construction, or steel feet set into foundations or footings, may be subject to considerable corrosion over the years from dampness inherent in such areas where tanking may have been omitted or become porous. Silver sand for rendering or in a concrete mix and containing a high proportion of china clay fragments and mica is particularly suspect. The effect of weathering is to wash out the china clay content causing the mica to flake away, and the surface of the render or concrete then becoming very porous causes the reinforcement under to rust and crack or even throw off the cover. Readers are recommended to study *BRE Digests* 263, 264 and 265 which deal with the protection, diagnosis and assessment of deterioration and repair of reinforcing steel in concrete.

In connection with concrete construction there have in recent times been a number of reported failures of concrete elements due to chemical attack or chemical deterioration. In April 1975 the Building Research Establishment (BRE) published a current paper titled 'High-alumina Cement Concrete in Buildings' (CP 34/75), in response to the collapse of two roof beams in a school in Stepney in February 1974.

The results of investigations published in the 1975 paper indicated that most HAC concrete prestressed beams used in buildings more than a few years old is highly converted and chemically changed or likely to become so. The strength of highly converted concrete is substantially reduced.

A considerable margin of safety is, however, provided for in most pre-cast concrete beamwork and slabwork. Proof loading tests in the field and experience in practice resulted in the BRE proposing that for all practical purposes the risk of failure of pre-cast beams or slabs using HAC concrete was very small for spans of up to 5 m.

Accordingly, if HAC concrete is suspected to have been used in a building with spans of 5 m or more, investigation should be made to:

(a) confirm if in fact HAC was used,
(b) determine whether the loss of strength due to chemical conversion may be significant for the strength of the structure.

Such an investigation could be undertaken using a simple pull-out test to assess the strength of *in-situ* concrete, as suggested in a BRE Current Paper published in June 1977 (CP 25/77).

There are several other ways in which concretes can suffer chemical attack or deterioration which could necessitate advising special tests. A concrete mix can also be less than ideal, or variable, and thereby causing a loss of strength. Good building practice is not always observed on-site, and concrete foundations laid in wet trenches or mixes which are incorrectly or unevenly proportioned can frequently be traced.

In the event of soil conditions indicating a danger of sulphate attack, special sulphate-resistant concretes should have been used but this is not always so and is generally unlikely in the foundations to older buildings. Deterioration is also possible if the concrete is in contact with other material which contains a high sulphate content, such as certain types of clinker breeze. Chemical attack from chlorides is also possible and occasionally encountered during a survey.

In addition to the possibility of chemical attack or chemical deterioration of the concrete itself, a possibility also arises of corrosion in the reinforcement. Both types of deterioration will be more likely and rapid in damp conditions, often being found to occur together for similar reasons. The amount of protection given to reinforcement in the past was frequently insufficient. Once the concrete cover protection given to the reinforcement begins to crack and soften, then wind and weather can begin to attack the reinforcement itself. Swimming pool roofs are a known problem area as they often have large spans of pre-cast concrete construction situated in a warm, humid atmosphere, and subject to high levels of condensation and chemical attack from chlorine. They may also become subject to weather penetration at defective jointing or on flat roof surfaces.

Before 1977 some concrete work incorporated calcium chloride as an additive to speed up the hardening process. This material absorbs water and forms a corrosive agent which rusts reinforcement bars and forces off the encasement; the combination of spalling encasement and rusting reinforcement may be highly damaging. Testing similar to that advised for HAC is required to confirm the presence or otherwise of this additive.

Occasionally it may be found that woodwool slabs have been used as permanent formwork to reinforced concrete, and this is a potential threat to the strength of the concrete because alkalis and sulphates which may be leached out of the slabs can react with the concrete; such reactions may be especially severe if HAC is present and in damp conditions. In addition, any permanent formwork may hide voids where the concrete has failed adequately to encase the reinforcement because of insufficient vibration during casting.

A significant proportion of commercial and institutional buildings erected since 1950 have been constructed with curtain walling using large glazed areas and spandrel panels; thermal and sound insulation is often poor and the glazed areas often face afternoon sun and prevailing weather so that deterioration to exterior joinery is severe. A large part of Britain's £20-billion stock of school buildings were erected using this type of construction and decaying timber joinery and plywood cladding used in such buildings has been reported as a major problem by 20 organisations surveyed by the Department of Education. Poorly selected softwoods was often used, with a high proportion of sapwood and rarely tenoned, dowelled or glued to restrain shrinkage; maintenance has often been poor. In some cases complete replacement of a facade will provide the best repair and offers an opportunity to improve thermal insulation and use better materials.

After reading the foregoing summary the reader may feel that reports on non-residential buildings present rather a formidable challenge to any surveyor's technical knowledge and to an extent this is true – certainly some additional research is required in many cases. The main points to bear in mind are the specific needs of the client and the purposes to which the building will be used; surveyors should always have at the backs of their minds the thoughts 'What are the possible pitfalls for this specific client in buying or renting this structure? Will he have allowed in his cash flow for the maintenance and other costs which will arise? What should he be warned about?'

13
Reports on flats and apartments

We hear every sound they make in the flat above; and some sounds they don't make as well.

Ground floor tenant

Frequently a surveyor's report is required on a flat, apartment, maisonette or other unit of accommodation which is part of a building divided laterally as well as vertically. Such property will normally be lease-hold, this factor alone requires special treatment (see Chapter 16). Very occasionally a 'freehold' flat or apartment will be encountered in England and Wales and the equivalent of a 'freehold' will be the rule rather than the exception in Scotland where different law applies.

If a freehold flat or apartment in England or Wales is to be surveyed, the client should be advised to discuss the implications of tenure with his solicitor. There are certain technical objections to 'flying freeholds' under English law as it now stands, because of the difficulty of enforcing positive covenants against the subsequent purchaser of a flying freehold who has not himself had any direct contractual relationship with the adjoining owners. This problem has been solved in England and Wales by the practice of granting long leases on flats and apartments rather than selling with a freehold interest. Commonly, 99-, 125- or 999-year terms will be encountered, the latter being an over-optimistic figure to apply to the life expectancy of the structure, especially in the case of some post-war speculative flat developments which will be fortunate if they see the end of a 99-year term! In addition to the lease the flat owners may also hold a share of the freehold.

There are two important points affecting flats and apartments:

1. the design of the building with particular reference to sound-proofing,
2. the means of escape in case of fire.

These are of profound importance to the client if he is to enjoy quiet occupation and physical safety. The survey report should always set out to cover sound-proofing as far as possible and if a noise problem seems likely, the client should be warned.

An elderly couple retiring to a ground floor apartment and hoping to end their days in tranquility will be bitterly disappointed to find that every sound made by a young family in the apartment above can be heard, perhaps late into the night.

Fire-proofing between units and the means of escape in the event of fire are vital matters. The standards to apply would be those of the current Building Regulations and also those which would be required before a fire safety certificate would be issued for a new building. If an existing building should fail to meet such standards, the client must be warned and recommendations made. Often, it would be impossible to meet modern requirements in an older building without the full cooperation of all flat owners, and this would be unrealistic. In such circumstances, all the surveyor can do is advise and warn and then leave it to the client to decide whether or not to proceed with the purchase.

Types of flat

In large conurbations, especially in London, there is a ready market for converted flats which began life as conventional houses or other structures and which have been subsequently sub-divided, often to a poor speculative standard. Caution must be exercised in reporting upon such property as there are several potential pitfalls, both physical and legal involved, about which the client will need to be advised.

Generally speaking, in England and Wales there are three main types of flat or apartment development:

1. Those in the form of simple low-rise structures, either purpose-built or converted, consisting mainly of two-storey buildings similar in appearance to two-storey dwelling-houses where separate units of accommodation are provided on each floor and where each unit has its own exterior entrance door. There may be a shared front porch or perhaps one doorway in the front elevation and one at the side. Such accommodation, in estate agents' parlance, is generally referred to as 'a pair of maisonettes' (in the London area at any event). Generally, garden areas are also divided and each unit, together with its garden is held under a separate long lease. Commonly each leaseholder is liable for the interior repair and decoration of his or her own unit. For carrying out repairs to the exterior or structure of the building, a variety of arrangements may be provided, these depending upon the date of the leases since conveyancing fashions have changed over the years, although generally speaking such arrangements fall into one of two broad types. Each lessee may be liable for his or her own part of the building, having no responsibility for any other part. A lateral division of responsibility is provided for the internal structure, this generally being the base of the first floor joists with a proviso that the staircase to the upper unit is included with that property. Alternatively, lessees may be jointly responsible for repairs to the foundations, exterior, structure and roof so that in the event of a repair being required, agreement would have to be reached on the need for and the cost of the repair. Each party would then contribute towards the cost in proportion to their holding.

2. Next is the modern purpose-built block of apartments, generally of post-war origin and with communal grounds, staircases and entrance halls. Occasionally services such as porterage, hot water and central heating are also provided on a communal basis. In such cases there will generally be a liability for each lessee to maintain the interior of his or her own unit and contribute a service charge for the maintenance and redecoration of the exterior, upkeep of the structure and roof and towards the communal services. Responsibility for arranging the communal services may vest in the freeholder who will then collect a ground rent, service charge and (usually) the insurance premiums due from the lessees, and arrange for the necessary periodic redecoration and upkeep of the exterior and structure of the building, its grounds and the communal services.

   Alternatively, the responsibility for maintenance could vest in a separate body which may be a residents' association with a chairman, secretary and treasurer, being subject to rules of procedure and with published annual accounts. Such an association will balance accounts by levying on members a service charge which is sufficient to cover the outgoings. The effectiveness of a residents' association will depend not so much upon the specific rules governing proceedings as on the quality of the residents who may be persuaded (or more usually cajoled) into serving. A well-run residents' association which maintains carefully the communal areas and grounds and ensures a high standard of maintenance and decoration of the buildings is the ideal. Occasionally, indifference on the part of residents or clashes of personality within an association will produce an unhappy effect, some members being anxious to maintain high standards and others clamouring to keep down costs. The standard of maintenance will then suffer considerably.

3. The third type is the older mansion block in larger cities, originally built for letting by private landowners or institutions and having been subsequently subject to 'break-up' operations involving the selling-off of individual units on long leases. Often, original tenants remain in such blocks so that a mixture of tenures is present, some residents having bought long leases and others occupying flats under regulated tenancies protected by the Rent Acts. In such cases either a freeholder will be liable for exterior and structural repairs with a right to recover service charges, or the freeholder would have created a residents' association of some type and be in the process of devolving responsibility for such matters.

Fig. 13.1 Flats system built in the 1960s using pre-cast concrete panel construction. This construction may be unacceptable as mortgage security for most mainstream mortgage lenders.

Service charges

In preparing a report on a flat or apartment held on a long lease, the surveyor should first read the lease (see Chapter 16), and if the lease provides for a service charge to be levied, he will wish to look with particular interest at the supporting provisions. There are four main requirements for a reasonable service charge provision, these being:

1. The period of payment should be given, that is, whether quarterly, half-yearly, yearly or on some other basis.
2. The items of maintenance, repair, decoration and upkeep, together with all services intended to be covered, should be set out clearly.
3. The manner in which service charges are apportioned between separate units should be fixed. Probably a number of identical flats in a block will pay the same amounts but flats of different sizes may need to be differentiated by, for example, council tax band, and other modifications may be needed in the interests of fairness (e.g. the ground floor lessees should not pay for lift costs).
4. The manner in which the maintenance accounts are to be kept and certified should be made clear.

Commonly, where a service charge is provided for in a lease, the freeholder will wish to ensure that all possible liability is developed upon the lessees and that no residual liability for extraordinary items of expenditure remains with the freeholder. As a consequence, those parts of the provisions dealing with requirement 2 above will often be very widely drawn and embrace all conceivable heads of expenditure. There may even be provision for additional matters to be added in the future.

There is a long history of actions in law and statutory intervention as a result of disputes which may arise between lessors or residents' associations on the one hand and individual lessees on the other over the reasonableness or otherwise of service charges. In *Finchbourne Ltd.* v. *Rodrigues* ([1976] 3 All ER 581) the Court of Appeal held that a term should be implied into a service charge clause that costs recoverable by the lessor from the lessee should be 'fair and reasonable' in the circumstances, and also that where the lessor and the managing agents for the property are effectively one and the same, the managing agents will not be sufficiently independent to certify the service charge under a provision for certification.

Section 91a of the Housing Finance Act 1972 was introduced in an attempt to assist lessees and was subsequently modified by Section 124 of the Housing Act 1974 and Schedule 19 to the Housing Act 1980. This provision provides that a service charge shall only be recoverable from the tenant of a flat for the provision of chargeable items to a reasonable standard in line with *Finchbourne Ltd.* v. *Rodrigues* (supra), and also that 'to the extent that the liability incurred or the amount defrayed by the landlord in respect of the provision of chargeable items is reasonable'. Provision is also made that in building works costing more than £250, at least two estimates must be obtained by the landlord, one at least of which is from a contractor wholly unconnected with the landlord or the managing agent. In respect of expenditure exceeding £2000, there is a procedure for the landlord to consult with the tenants except in cases of emergency and only then to obtain estimates for the work.

The provisions of the Housing Act are designed to enable tenants to obtain information about service charges even if their leases conferred no such right. It may well be that in order to complete his report that surveyor will need to inspect the management accounts. It is unusual to find that a prospective purchaser or his surveyor or even his solicitor have a statutory right to do this but the vendor of a long lease will have such a right, either by virtue of the provisions of the lease itself or by virtue of Section 90. If the surveyor is denied access to this information by the vendor, or if it is suggested that up-to-date accounts are not available, then this must raise serious doubts about the present state of management accounts and the matter may need to be referred back to the client's solicitors for further instructions.

If an inspection of the management accounts has not been possible, the surveyor will be unable to express any view on the reasonableness (or otherwise) of the service charge currently payable, or the likelihood that a substantial future increase in the service charge may be possible because of past under-provision for some items.

When an inspection can be made this should preferably be of accounts over a period of more than 1 year. Ideally, the last 3 years' accounts should be scrutinised. In order to forecast future trends, these accounts should be considered in the light of the need to maintain some reserve for periodic decorations, often necessary at 3-yearly intervals, this causing a sudden leap in the service charge for 1 year if an adequate float has not been built-up in the two proceeding years. Similarly, special attention should be given to checking the provision for capital replacement of items such as lifts and fixed plant which may have a well-defined lifespan. If there are fixed-term contracts for cleaning or gardening and if these are about to expire, a large increase in such costs may arise if an alternative contractor has to be found.

An inspection of a single flat in a block of flats for a prospective purchaser must, of necessity, include not only the flat in question but also the building as a whole since a sudden large increase in the service charge could arise in respect of a defect in the building quite unconnected with the flat under review. In my experience this will often arise in the case of major structural defects to foundations and roofs. It is essential for the client to realise that he may be proportionally liable for this type of defect even if it is not related to the flat he proposes to buy. Similarly, a prospective mortgagee will wish to know about such matters and will expect the building as a whole to have been inspected by the surveyor in addition to the specific flat used as security.

The Royal Institute of Chartered Surveyors (RICS) Practice Note, 'Structural Surveys of Residential Property' states that 'Particularly onerous repair liabilities may exist quite independently of the subject property, such as where a lease or agreement imposes a liability to pay a proportion of the total estate repair costs', and 'To advise fully the surveyor must endeavour to establish the extent and nature of the repairing liabilities under the lease, the client's potential responsibility for executing repairs and also his liability to pay for repairs executed by others'.

If a block of flats suffers from a serious structural defect, the purchase of one of the flats, even where unrelated to the defective area, may be unwise unless all the implications are fully understood. It may be, for example, that a foundation defect exists and that this should be remedied by the National House-Building Council (NHBC) Certificate procedure, or works may be covered partly by the subsidence and landslip provisions in the building's policy. In such a situation, there may be a number of parties involved in a liability dispute including the lessees and their solicitors, the original contractor, the NHBC, the building insurers (possibly under individual policies on the flats) and the local authority who passed the work originally. There may also be a residents' association, and an owner of adjoining trees, the roots of which could have contributed to the defect.

All surveyors will be aware of problems associated with flat roofs, especially those of poor construction dating from the 1950s and 1960s. Many such roofs were used on blocks of owner-occupied flats and the roof coverings are now nearing the end of their useful lifespan. Renewal will often be a question not merely of stripping off existing built-up felt or asphalt and re-surfacing. Frequently it will also involve repairs to the decking and supporting construction involving major expenditure on the works and on protection of property and occupiers for the duration of the works.

A major item of expenditure such as renewal of a flat roof covering, presents particular problems to a residents' association or freeholder collecting periodic service charges. Ideally, a sinking fund should have accumulated to pay for this work but this will rarely be found to be so. It may be that the nature of the remedial works required will have to be agreed (itself not an easy matter) and then estimates obtained and contractors instructed on the basis that a sudden large increase in the service charge will have to be made to cover this item.

The chairman of a residents' association in such a situation will feel like the messengers of old who were praised if they brought good news but put to death if the news was bad. The owner of a flat in such a block may actually decide to sell and move just to avoid problems of this type which may be approaching and the purchaser will need to be warned about such circumstances.

Tenure in Scotland compared

In Scotland a different law applies to the sale of flats and apartments. Scottish law is based upon the original feudal tenure modified by statutes, with all rights to land originally derived from the Crown. Buildings and landed property are classified as 'heritable property' and being either 'corporeal' if land or buildings, or 'incorporeal' if merely rights of land or buildings. In these respects the system is similar to that operating in England and Wales. One difference is that a system of charges may be levied, similar to rent charges on freehold land in England and Wales, and termed 'feu duty', normally running in perpetuity with the land, the duty being paid by the owner or 'vassal' to the person or 'superior' who owns the right to receive the duty. Occasionally a superior may have a right to acquire the vassal's interest by virtue of a clause in the title which, in effect, gives him an option to purchase, generally at a price equal to any other offer which the vassal may have received for his interest. Such a right to pre-emption normally arises only on the first sale of the vassal's interest and not subsequently. A similar situation arises in England where an original freeholder may impose restrictive covenants on subsequent purchasers of the freehold interest, although a right to buy back the freehold is less usual.

A wide range of feuing conditions may be imposed on vassals by the superior and may generally be enforced, not only by the superior but also by other vassals who hold parts of the same original holding. This is similar in many respects to the estate covenants which may be imposed by freeholders of estates in England and Wales whereby a series of mutually enforceable covenants may apply to the estate as a whole. This is a particular feature of some residential developments built between the wars in the higher price ranges where obligations may arise to maintain private roads, open spaces and the like with limitations on the type of alterations or uses permitted.

In English law there must be privity of contract between two parties before action may be taken, but in Scotland a vassal or 'feuer' may take action against another without there being privity of contract between them provided that the aggrieved feuer is personally affected by the breach of feuing conditions and that the superior has not himself permitted the breach specifically.

Against this background of mutually enforceable feuing conditions, which arise irrespective of whether or not there is a direct contractual relationship, it has been possible for property in Scotland to be sold as 'flying freehold' subject to the feu duty and to the conditions. Scottish law also permits servitudes (the equivalent of English easements) in the form of rights of support, protection and light.

During the course of inspecting a flat or apartment in Scotland the surveyor will need to obtain details of the feuing conditions set out in the title and consider whether or not these are adequate and suitable. Insofar as potential liabilities may arise for the maintenance and repair of either the structure as a whole or other parts of the building separate from the specific flat concerned, these may need to be investigated and the position explained to the client.

Conversions

Apart from general distinctions of tenure and various types of flats and apartments which may be encountered, there are two broad divisions of type which need to be identified. These are: (a) those flats or apartments which have been purpose-built and (b) those which are conversions. The surveyor should make a particular point of confirming in all cases which of these two situations applies since this is relevant to the advice which may be given and may be significant to a prospective mortgagee or client proposing to raise a mortgage. Institutional lending policy on converted flats is often different from that applied to purpose-built flats.

In most cases the converted flat will be fairly obvious. An original Victorian or Edwardian villa in a city suburb may have two entrance doors in the porchway and a plasterboard and stud partition wall dividing the original hall. The nature of the original house and the extent of the conversion works will be self-evident in such circumstances.

Faced with such a conversion, the surveyor should see if the staircase to the upper flat has the necessary degree of fire-resistance and that the partitioning and undersides of the staircase are suitably fire-resistant, at least up to the standards of the current Building Regulations or London Building Acts. Similarly, if the upper floors are simply suspended timber floors, these will have required additional fire-proofing to meet regulations. The reason is to ensure that a fire in the lower flat will not spread rapidly to the flat above and that in the event of fire, the occupants of the flat above may have a protected means of escape down the stairs.

Many converted flats have never been subject to any scrutiny by a building control officer or district surveyor. The reason for this is that they have been 'converted by stealth'. Probably a house was first used as accommodation for two families. Later, various modifications were perhaps undertaken, culminating in sub-division of the accommodation and splitting of the services. If the change of use to two units of accommodation is of long-standing and if internal alterations are purely internal, no reference to the local planning department may have been made. Finally, the tenants leave and the owner, much impressed by the current level of property values for converted flats, will decide to sell the house in two parts by creating two separate leases. At this stage, one of the purchasers commissions a survey and finds that there are wide-scale breaches of the Building Regulations or other regulations. If the fire-proofing between units is affected, the building may actually be dangerous. The surveyor will often find partitions of plywood or hardboard masquerading as plasterboard. Often various defects such as rising damp, decay to timbers, defective ceilings and roof problems will have been dealt with in a highly unsatisfactory manner and a variety of defects may have been plastered and decorated over.

Occasionally it may be difficult to confirm initially whether or not a particular property has been converted, and some enquiries regarding its history or exposure of the structure could reveal the truth.

Rights to manage, extend leases or buy freeholds

In England and Wales we now have further legislation which extends the rights of leaseholders of flats in many situations. Leaseholders may club together to buy the freehold, or may take over the management of blocks when existing management is poor. Individual leaseholders may apply for a lease extension if the shortening lease term is affecting value or saleability of the flat. As is so often the case with government tinkering with the leasehold system the consequence is an additional level of complexity and plenty of scope for further litigation.

14
New buildings and buildings under construction

Where there is a contract to build and sell a house or to sell a partly-built house when it is completed, the law makes a three-fold implication: that the builder will do his work in a good and workmanlike manner; that he will supply good and proper materials; and that it will be reasonably fit for human habitation.

Hancock *v.* B. W. Brazier *(Anerley) Ltd (1966)*

Horses, carts and the price of land

Surveyors often have to inspect building sites in the process of development or brand new completed buildings. The report may be for the buyer and may, or may not, include some advice on value, or it could be that a report is being prepared for a mortgage lender perhaps in addition to a report for the purchaser. So we are involved here with the process of property development.

I was at a meeting recently when a spokesman for one of the volume house builders was complaining about the high cost of land. He said 'The high cost of building land is driving many first time buyers out of the market because it is making new homes too expensive'. At first glance this may seen a very reasonable proposition. Of course, one might think, if land values rise this must push up the price of houses. And house builders are always pressing for the release of more land for housing in those places where people want to live, especially the greenfield sites in the south-east of England.

But is it the high cost of land that pushes up house prices?

The prices builders charge for their house are not, in fact, determined by the cost of land. In specific cases there is no relationship at all between the two. If a builder bought a site cheaply 20 years ago and has kept it in his land bank ever since, this will make no difference to the prices of the houses he will build; he will still charge the highest price that he thinks the local market can stand.

Likewise if a builder pays a silly price for a site, far in excess of its real value, this will not mean that the houses he builds will have a higher value. If he over-prices the houses they will not sell. The value of the finished product is determined by market forces and at the end of the day the completed development can only be sold on terms dictated by the market.

So the builders' spokesman was putting the cart before the horse; it is the high selling prices of houses that cause land values to rise. An artificially low land price would simply result in the developer making a larger profit. If houses were sold at an under-value the net effect would be that purchasers would buy on an estate and then resell at a profit. Whatever happens the new houses would soon find their level of value in relation to other houses in the locality.

A residual valuation of a site takes as its starting point the value of the completed development and from this is deducted the development costs including building costs, rolled up interest charges and professional fees. A further deduction is made for the developers' profit (varying with the degree of risk and likely competition for the site).

Residual valuations are something of a minefield for the unwary. There are so many variables and building costs are notoriously difficult to estimate accurately without full plans and a bill of qualities. So the final selling price of the site will be determined by competition in the market from a number of developers, each of whom will have prepared his own residual valuation. The site will sell in the end to the developer who is most optimistic regarding the completed value of the development and whose calculations show the most cost effective development.

The traditional residual valuation is now much refined and computer models are used. The old style developers who flew by the seat of their pants and worked out what they could pay on the back of an envelope have had to adapt to this or fall by the wayside.

So the house builders know the score. Before bidding for a site they do their sums. And they start off with some market research to find out how much they can sell the houses for.

A number of bodies in the UK provide warranty schemes for new homes including Zurich and the National House-Building Council (NHBC). The NHBC publishes a most useful guide to its standards and requirements – the National House-Builders Handbook – and this is essential reading for any surveyor reporting on new houses or flats. The Handbook together with updates and other NHBC publications provide guidance on NHBC standards and some useful insights into the problems most likely to arise with new homes and the issues most likely to upset purchasers.

When undertaking a Building Survey on a new home, or a house under construction, the procedures for taking instructions, undertaking the inspection and preparing the report will be the same as for older buildings but certain special considerations apply to new buildings and these need to be reflected in the investigations made and the topics covered in the report.

These special considerations include:

1. The terms of the contract between the builder or developer and the client.
2. The statutory and other consents required especially Building Regulations.
3. The fact that the foundations and structural design of a new building cannot be seen to have proved themselves by the passage of time.

The contract

The client will have a contract with the builder or developer and will need to know if the terms of that contract have been fulfilled. So the surveyor will need to see the contract and will need to know if, as part of the terms, a 10-year NHBC, Zurich or similar certificate is being issued, in which case the construction will need to comply with the requirements of such body.

The certifying bodies carry out spot checks on site but their inspectors cannot be there all the time, so poor construction practices and breaches of their standards may occur. The common law position is that in the absence of any specific contractural terms the builder must do his work in a good and workmanlike manner, supplying good and proper materials, and the finished property must be reasonably fit for human habitation.

The surveyor will need to see the plans and specification, if these form part of the contract, and will need to compare these with what has actually been built. Of course, the contract may provide that the builder can vary the specification and sometimes the contracts used for new house construction are amazingly woolly giving the developer freedom to make significant alterations. At the end of the day, however, the client needs some reassurance that he or she is getting what they thought they were getting when they paid their deposit.

Planning and other permissions

Another important point to consider in any new building is that almost all new construction in the UK is subject to the requirements of various permissions and consents and buildings erected without all the necessary permissions and consents may be illegal or unlawful. Clearly, before a client buys or leases a new building and before a mortgagee lends money on the security of a new building, the fact that all necessary consents have been obtained must be confirmed.

In the first instance, it is the client's solicitor who is responsible for finding out whether or not the new building exists legally and lawfully and that there are no breaches of the law. As has been mentioned before, however, it is the surveyor who actually looks at the property – the solicitor will rarely see the property he conveys for a client so the advice which the surveyor can give his client or his client's solicitor will prove to be important.

Basically, there are four types of consent which may be required for new buildings:

1. Permissions under the Town and Country Planning Acts, except for 'Permitted Development' in the case of certain types of small extensions to existing buildings or to exempt buildings such as those used for agricultural purposes.
2. Permission under building control codes, quite a separate system from planning and governed by the Building Regulations 1985, which were amended as from 1 April, 1990 in respect of approved documents F, G, H, J and L and which provide the basis of building control in England and Wales. Similar regulations apply in Scotland and Northern Ireland. Building construction legislation is subject to periodic review, and it is important to keep up to date.
3. The consent of a landlord or freeholder of leasehold property; of an original freeholder in England and Wales who holds the benefit of restrictive covenants; or, in Scotland, the superior to a title who holds the benefit of feuing conditions. Examples may be found where titles include old restrictive covenants against redevelopment at higher densities or different types and such conditions may be removed only by applications to the Courts or Lands Tribunal. Specific insurance cover may sometimes be available against the possibility that an old restrictive covenant could be enforced and building works may then proceed with the benefit of such insurance cover. If title insurance is relied upon, the surveyor should obtain details and confirm whether or not the amount of cover is adequate since it will often be found insufficient to compensate fully a purchaser in the event of a restrictive covenant being enforced.
4. The consent specific to the particular type of building having regard to particular local or national legislation. Special rules apply to filling stations and garages which require petroleum licenses and if a petroleum licence is refused or revoked, this could have a disastrous effect on the value of the premises. Similarly, fire safety certificates are required for hotels, public buildings and the like. It is important to confirm that a new building meets all such requirements, including projected changes of legislation which may arise in the future. Offices and shops are governed by the Offices, Shops and Railway Premises Act and factories by the Factories Act. In the case of buildings where people will be employed, the Health and Safety at Work legislation may apply to certain aspects of new building design.

If a new building is not residential then the surveyor will consider whether or not the design complies with all existing and projected local and national legislation to which it will be subject, having regard to the purpose for which the building was designed and the purpose for which the client proposes to use it, taking into account any potential change of use.

The fact that planning permission and approvals necessary under building control legislation have been granted does not ensure good design. The fact that a new building may have been inspected periodically by a building inspector or district surveyor does not ensure good standards of construction since such inspections, by their very nature, can only be 'spot checks'. Neither will the fact that a new house may have been inspected by the NHBC indicate good standards of construction, or even that all NHBC requirements have been met, since these inspections also are only spot checks.

Therefore, when reporting on a new building, the surveyor should not accept anything at face value, nor rely upon the past supervision or inspection by others as an indication that a particular item is satisfactory. A surveyor can only report properly on those matters which can be seen or discovered in person.

Foundation design and subsoil

The first subject for the surveyor's investigation is the nature of the supporting subsoil and the foundation design. In this connection, the architect's or builder's design drawings as approved by the local authority will at least show the intentions. Consideration should then be given to the question of whether or not a design is suitable for the indicated subsoil conditions and enquiries made to discover what subsoil samples were taken and whether or not any trial bores were made on the site and to what depths. It may be that results of a laboratory examination of soil samples can be made available. If not, and if the position seems doubtful, it may be that the surveyor will need to recommend that his or her client commissions an analysis of the subsoil. In recent years, however, most local authority building control departments have required a note to be made on the submitted drawings to the effect that the foundations will be designed to meet site conditions rather than requiring an explicit detail on the drawings.

The reason for this emphasis on the foundations and subsoil under a new building is that if a building has not proved itself by performance over any extended period of time, the danger, albeit remote in most cases, is that the foundations may be inadequate or the subsoil unstable. From a client's point of view, the whole point of instructing a surveyor in the first instance is to receive advice regarding just such matters and a report which merely states that the building has not been subjected to any extended period of observation and that no advice can be given on the suitability of the foundations will seem distinctly unhelpful to a client. He will then wonder whether the exercise of instructing a surveyor was really worthwhile.

Certain obvious warning signs which should place a surveyor on his guard over a foundation design would include the following:

1. Any known history of mining in the locality.
2. Any known history of sand, gravel or chalk extraction or filled gravel pits, etc. in the locality.
3. Sites with steep slopes, retained by embankments or retaining walls, or where adjoining structures or highways show signs of landslips.
4. Any indications of tipping or landfilling in the locality.
5. The presence of shrinkable soils, especially near vegetation.
6. Any possibility of coastal erosion.
7. Signs of structural distress to older buildings in the locality.
8. Trees closer to the structure than their mature height and which could cause subsoil shrinkage. Evidence of the recent removal of trees from a site where ground heave is likely.
9. Adjacent flat, marshy or poorly drained land.

Another useful indicator is to make a check on the identity and qualifications of those responsible for the foundation design. Generally, it is to be hoped that consulting engineers were employed to report on soil conditions, prepare structural calculations and designs and that qualified architects or surveyors were responsible for working drawings and layouts. Often, however, drawings are prepared by 'inhouse' personnel employed by the builder or developer.

Occasionally drawings will be found carrying no indication as to the name or qualifications of the draughtsman. No hard and fast rules can be given for the acceptability or otherwise of designs prepared by unqualified or unknown personnel since they may well be experienced and competent. However, the lack of any outside professional involvement in structural design can be an indication that it was felt necessary to avoid payment of professional fees and cost cutting may then also be evident in the design or the construction work itself.

It is sometimes helpful to talk to the men working on a building site if this is possible and it is surprising how much can be learned from such casual conversations. They will give an indication of the loyalty of the workforce and the efficiency of the employer in providing adequate supervision. They will often readily discuss problems which they have encountered on site, difficulties with the local authority or faults and poor deliveries of materials. An overall impression of the general standard of care being taken with the building work can be gauged. On no account, however, should members of a workforce be 'quizzed' on the progress, administration and finances of a contract.

Inspecting the building

Having completed all the investigations which are possible and are required by the client in respect of subsoil and foundation design, the surveyor will then proceed with the inspection of the building. This inspection will follow the same lines as the inspection of any other building using the surveyor's own normal routine. If the building work is incomplete, it is suggested that a list of all unfinished items be prepared and included in an appendix to the report with a note that the surveyor, if required, can confirm that such matters have been completed at a later time.

At each stage of the inspection, all defective work and deviations from good building practice should be noted. In the case of a residential property a check should be made for breaches of building control legislation applicable to the locality and any lack of conformity with the standards laid down in the NHBC's *Registered House-Builders' Handbook*. In the case of non-residential property, the check should be for breaches of building control legislation and any other lapse from the specific requirements of that particular building type.

Surface finishes must be checked with particular care, especially in the case of a residential property. Clients buying new houses are doing so for many reasons, but one requirement will be that the decorations will all be new. The clients will then expect that redecoration will not be necessary for several years, this clearly being a saving compared with the typical situation in a second-hand house where most buyers expect to have to redecorate in accordance with their own tastes soon, if not immediately, after moving in. Experience shows that purchasers of new houses tend to complain most over problems of final finishes to walls, ceilings and woodwork and by lack of care when installing fixtures and fittings. It is necessary, therefore, to apply a stringent standard to a new house where these matters are concerned, as compared with a second-hand house where most clients will accept a lower standard knowing it is their intention to improve to their own tastes. Having said this, however, clients should be told that a builder's initial finish is not as durable as later-applied finishes may be, especially in the case of new painted softwood.

During the course of a new building inspection, an opportunity arises for the surveyor to consider the overall quality of design from the client's point of view and to comment unfavourably where necessary. In commercial and industrial structures the durability of finishes is important in view of the considerable wear and tear which may arise in connection with the client's trade or business. Floor finishes in commercial and industrial structures are very important from the point of view of hardness, freedom from dust, ease of cleaning and non-slip surface. Often, hazardous floor finishes or potential dangers in a building will be encountered which may possible give rise to claims from injured employees or visitors. Clients will expect to have been warned of such problems by their surveyor in advance.

Security in a new commercial or industrial building may be poor and a client may have to undertake protective improvements before being able to obtain insurance cover, especially if warehousing is involved or valuable goods stored. Often new factory and warehouse space is built and let, or sold, as a shell on the basis that the purchaser or tenant will undertake his own fitting-out. The provision of adequate security, fire safety precautions and fitting-out to comply with local and national legislation will have to be allowed for by the client and the advice of the surveyor in budgeting for these costs may be relevant.

The quality of fittings provided in new speculative buildings may be poor. In dwelling construction there is a wide range of qualities available in items such as sanitary ware, kitchen units, fenestration and door and window furniture. The temptation for the builder or developer is to use fittings which are superficially attractive but less durable, especially in the case of sanitary ware. A client may see a new and attractive bathroom and be impressed by the appearance of the sanitary ware and fittings, without being aware of the distinction between acrylic, steel and cast-iron baths, or the variations in enamel thickness, or the tendency for acrylic to flex in use and suffer surface deterioration from abrasion. At the present time, for example, it is possible to buy baths at prices which, for the most expensive, are 10 times the cost of the cheapest with very little difference in superficial appearance.

The internal planning of new accommodation is important. In residential construction it is necessary to consider whether or not the rooms will be of suitable shape and size to accommodate normal furniture satisfactorily and if furniture can be conveniently moved in and out. For example, to be usable bedrooms must be of a size and shape to accommodate a bed while leaving adequate access for bed-making and space for the door to be opened. (This may seem obvious, but there are many instances of small bedrooms being virtually unusable for this reason.)

Suitable access for reading electricity and gas meters and provision for the storage of rubbish bins are necessary. External meter boxes are clearly preferable and should be located in front of any lockable side gates so that access from the highway is always free. Similarly, rubbish bins should be kept out of sight but convenient for use and for emptying. Correct planning of water closets and bathrooms is necessary to ensure adequate room for the convenient use of appliances and door opening. Convenient kitchen layouts must avoid excessive movement being necessary from one area to another in kitchen operations; a layout of worktop – cooker – worktop – sink – worktop – fridge/freezer is preferable, unbroken by doorways or trafficways.

Special attention should be given to staircase and balustrade design, this often being poor. Widths should be sufficient to allow for furniture removal; handrails should be adequate and extend for the full length of the flight; good lighting is necessary, and balustrades should be provided with gaps such as to prevent the smallest child from falling through or becoming jammed. The Building Regulations demand that the space between balusters should be such as to prevent the passage of a 100 mm diameter sphere.

Recent evidence tends to suggest that the overall quality of new building design and construction has improved during the 1990s and that lessons have been learnt from the failures of the previous two decades but there has also been a tendency to build on low-lying land subject to flooding and urban 'brownfield' sites subject to subsidence problems.

15
Reports on older buildings

I think that your client needs a psychiatrist rather than a surveyor if he seriously intends to buy this house.

Surveyor's initial reaction to instructing solicitor

In the end we are all dust, and our buildings also. Nothing – not even the great monuments of past civilisations – lasts forever. So in treating older buildings to a separate chapter the author wants to focus on the two aspects of most concern to surveyors in dealing with structures which may have passed through many ownerships and which may have been subjected to neglect or inappropriate building works in the past and which may be past their prime.

First, the surveyor has to be able to date the building. Knowledge of the period in which it was constructed is essential for soundly based advice to be given. A report on a house built in 1850 will be very different from that on a house built in 1950 – or indeed 1650. We tend to date buildings by reference to the kings and queens of England (even the Americans refer to Georgian, Victorian and Edwardian buildings) so on page 145 there is a schedule of the dates of our rulers and certain other important landmarks, such as the Great Fire of London in 1666.

Secondly, the surveyor has to be aware of local and regional characteristics which determined the choice of building materials used and construction techniques employed. Local knowledge can be vital. A surveyor who strays too far away from home looking at old buildings in unfamiliar parts of the country – or other countries – must take care.

The history

It is most unlikely that any surveyor will be required to report on a building constructed before 1066 during the Saxon and Danish periods except perhaps those specialists dealing with ancient monuments or ecclesiastical buildings who might occasionally look at some Saxon stonework. Houses from this period were of timber construction and have long since rotted away or burnt down.

Construction methods used before 1066 for dwelling houses continued long afterwards, however, in the form of cob, wattle and daub and thatch.

Stone was the principal material used for church and cathedral construction and this was difficult and expensive to transport by horse and cart, river or sea. From the earliest times local materials have been used from as close to the site as possible and we expect to find local stone used for all but the most expensive buildings during the Norman period. The limestones of Portland, granites of Aberdeen and Cornwall, and Bath stone from the west country were all first used locally, and then as transport became more efficient and the country more prosperous stone from further afield began to be used for the more important buildings, as with the Normandy Caen stone used in Canterbury Cathedral.

For less grand buildings local materials have been used right up to the 19th century. The coming of the canals, followed by the railways eventually made it economic for brick, slate and tile to be transported more widely so that Welsh slate would be used to cover London's Victorian terraces.

Modern bricks came into use in about AD 1300, having lapsed since Roman times, and the earliest brick building which survived into the 20th century is Little Wenham Hall in Suffolk (AD 1260). The predominance of brick for domestic construction dates from the reign of William and Mary and it remains the most economic and universally acceptable material for the external facing to residential buildings.

Half-timbering as in Lancashire, Cheshire and the West Midlands dates mainly from the 14th to 16th centuries. Terracotta was used in Essex in the early 16th century and in localities with chalky soils flint was used and is still common in Norfolk, Suffolk and Hertfordshire.

In these areas the flints are split (or knapped) to provide a flat face and the wall is built using flints as random rubble set in lime mortar with the knapped faces providing an attractive weathering surface. Lots of churches in Hertfordshire and East Anglia together with village houses are built this way. The flints are extracted from gravel pits or harvested from the surfaces of the fields after ploughing, and is good cheap local material, easy to collect and transport.

In both ecclesiastical and secular buildings the earliest periods likely to be encountered are:

1189 to 1272 Early English
1307 to 1377 Decorated
1377 to 1558 Perpendicular

These three periods feature the Gothic style in Britain which survives generally in the form of ecclesiastical buildings.

The Renaissance period commended at the start of the 16th century after the Wars of the Roses and during the suppression of the monasteries (1536–1540). During this period we saw the conversion of some ecclesiastical buildings into mansions, the endowing and building of grammar schools, and during the Elizabethan period a great increase in the building of country houses such as Longleat in Wiltshire and Haddon Hall in Derbyshire.

The Jacobean period saw such mansions as Holland House in Kensington and Bolsover Castle in Debyshire built. These and the preceding Elizabethan mansions were principally of brick and stone but timber construction was still employed in Cheshire, Lancashire and Shropshire. Emmanual College Cambridge and Jesus College Oxford were built during this period.

The Anglo-Italian period began during the reign of Charles I who employed Inigo Jones (1573–1653) to design complex structures of classic grandeur. The Great Fire of London (1666) presented an

Dates and periods

43 to 450	Roman Britain
450 to 613	Coming of the English
613 to 1017	Division into Kingdoms
1017 to 1066	Saxon and Danish
1066	Saxon techniques replaced by Norman
1066	William I
1087	William II
1100	Henry I
1135	Stephen

(Continued)

Dates and periods *(Continued)*

1154	Henry II		
1189	Richard I	1189 to 1272 Early English	⎫
1199	John		⎬ Gothic period
1216	Henry III		
1272	Edward I		
1307	Edward II	1307 to 1377 Decorated	
1327	Edward III		
1377	Richard II	1377 to 1558 Perpendicular	
1399	Henry IV		⎭
1413	Henry V		
1422	Henry VI		
1461	Edward IV		
1483	Edward V		
1483	Richard III		
1485	Henry VII		
1509	Henry VIII		
1547	Edward VI		
1553	Mary I		
1558	Elizabeth I	1558 to 1603 Elizabethan	
1603	James I	1603 to 1685 Jacobean	
1625	Charles I		
1649	Commonwealth		
1660	Charles II		
1660	Restoration		
1666	Great Fire of London		
1685	James II		
1689	William III and Mary II		
1702	Anne		
1714	George I	1714 to 1837 generally referred to as Georgian	
1727	George II		
1760	George III		
1820	George IV	1810 to 1820 also as Regency	
1830	William IV		
1837	Victoria	1837 to 1901 generally referred to as Victorian	
1901	Edward VII	1902 to 1914 generally referred to as Edwardian	
1910	George V		
1914 to 1918	Virtually no construction during First World War		
1936	Edward VIII		
1936	George VI		
1939 to 1945	Virtually no construction during Second World War		
1952	Elizabeth II		

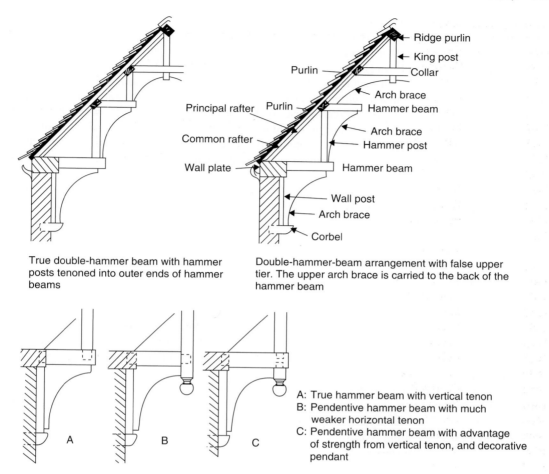

Fig. 15.1 Typical hammer-beam construction used mainly in churches and cathedrals. Detailed knowledge of this type of construction is essential before an accurate assessment of strength may be made.

opportunity for the redesigning of the city and the genius for construction of Sir Christopher Wren (1662–1723).

From here there was a logical progression of construction methods and styles to the 18th century which is split into two periods: Queen Anne (1702–1714) and Georgian (1714–1837) including the period of the Regency (1810–1820). Dwelling houses from the Queen Anne period were typically fairly to plain and functional but in well-proportioned facing brickwork and now much appreciated and sought after. Castle Howard and Blenheim Palace were designed during the Queen Anne period in the Anglo-Classical style whilst during the Georgian period Robert Adam (1728–1792) was probably the most famous exponent of the Georgian style (witness the Adelphi Terrace in London and Keddleston Hall, Debyshire). Columns, vaulting and sprawling plans again became fashionable with a strong emphasis on facades. This prompted the classical revival with ambitious pseudo-Greek models such as the Bank of England (1788) and St George's Hall, Liverpool by H. L. Elmes (1815–1847).

An upsurge in church building in the 19th century brought us the Gothic Revival with Augustus W. N. Pugin (1812–1852) as the apostle of ecclesiastical work in this style. The most impressive buildings of this time are, however, not churches but the Houses of Parliament designed by Sir Charles Barry with Pugin's help and the New Law Courts designed by G. E. Street (1824–1881), who was a pupil of the famed Sir Gilbert Scott. Further notable buildings of this period include Keble College, Oxford, Truro Cathedral and Manchester Town Hall.

Towards the end of the 19th century we had the Arts and Crafts movement, which influenced the design of small-scale residential development including the furniture and decoration. A good example of this is the Bedford Park Estate in Chiswick, West London, designed by Norman Shaw as an enclave for artists and writers where the distinctive style of the houses with brick and hanging tile elevations, balconies and fenestration are much appreciated now and where – at the time of writing – good examples sell for £1 million or more to the well-heeled London house-hunter looking for a characterful family home.

This then brings us up to the 20th century and the substantial and well founded works of architects such as Sir Edwin Lutyens RA whose Castle Drogo in Devon, completed in 1930, was the last great country house to be built in Britain.

The defects

Elizabethan and Jacobean

Generally walls are not plumb and vary in construction from timber lath and plaster via porous brickwork and eroded freestone to disintegrating cob. Roof timbers typically include cruck from which the bark has not been stripped and will be heavily infested with furniture beetle. Lots of oak and other hardwood used in the construction will be in contact with damp masonry and thus prone to fungal attack and deathwatch beetle.

Floors consist of stone flags or herringbone brickwork laid directly on the earth, with both floors and the base of walls prone to rising damp and damp penetration. In the case of cob walled cottages, however, the dampness is necessary, for if the walls dry out they disintegrate to dust, having then no moisture content to make the cob filling cohesive.

Queen Anne, Georgian and Regency

These are attractive in appearance, especially the front elevations, but the facade generally hides a multitude of defects. Walls may be in brickwork using a soft lime mortar with a considerable amount of timber built into the walls in the form of bressumers, lintels, wall plates and especially bonding timbers. Bonding timbers, usually 100×50 mm softwood, were built into the walls replacing a whole course of brickwork on the internal face and intended to add strength and stiffness. Walls subject to damp penetration from leaking parapets and the like will often show signs of alarming bulging and leaning due to rot and beetle attack to built-in timbers. The remedy, more often than not, is complete rebuilding of the wall.

Solid walls are frequently not thick enough to keep out the weather, for example, the 150 mm ashlar stonework used in the bath area for classical terraced housing. The stonework is often battened and lined internally and covered with lath and plaster or other material to cover up the damp.

The outside space to basement houses below pavement level is known as 'the area' and usually gives access to storage vaults (originally for coal) under the pavement.

Watch out for basements and ground floors with timbers laid directly on the earth and a general absence of damp-proof courses (dpc) or damp-proof membranes and inadequate ventilation, making an ideal breeding ground for wood boring weevil, furniture beetle and both wet- and dry-rot.

Traditional queen post and king post roof trusses may be found in this period with roof timbers generally better than in the earlier Elizabethan and Jacobean buildings.

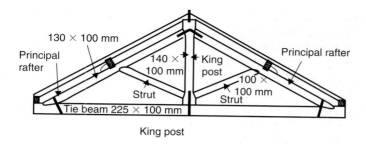

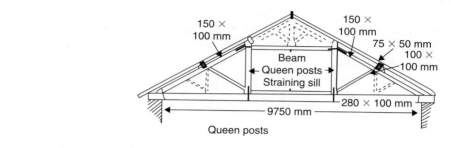

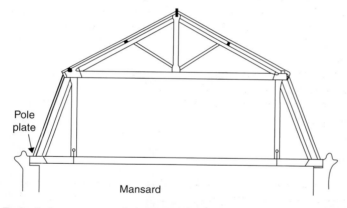

Fig. 15.2 Early timber trusses, variations of which will be commonly encountered in older buildings.

On the following page there are examples of designs for Rate Houses from London typical of the period.

Following the Great Fire of London in 1666 a Building Act was passed by Parliament in 1667, followed by another in 1774 specifying four sizes of houses permitted to be built in towns and cities: First Rate and its accompanying mews, Second Rate, Third Rate and Fourth Rate.

This standardisation of style and height gave rise to the homogeneity of Georgian London and the house types continued to be used through the Victorian period – with embellishments – as may be seen in South Kensington.

Victorian and Edwardian

Too much fussy detail to elevations and roofs gives rise to damp penetration and resulting timber defects. Parapet gutters often become blocked at the outlet, especially if there are trees nearby dropping leaves in

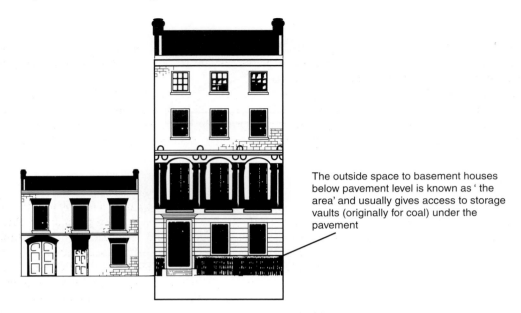

The outside space to basement houses below pavement level is known as ' the area' and usually gives access to storage vaults (originally for coal) under the pavement

Fig. 15.3 First Rate house and its mews.

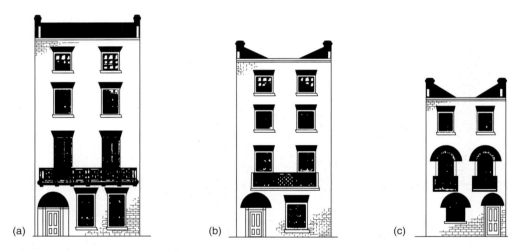

(a)   (b)   (c)

Fig. 15.4 (a) Second Rate, (b) Third Rate and (c) Fourth Rate.

autumn. Typical roof layout are either the centre valley (or butterfly) type with parapet at the front draining to a hopper head at the rear, or with a central ridge draining to parapets at front or sides. Roof timbers beneath the parapet gutters and centre valleys should be carefully checked during survey since these are invariably problem areas. A variation in some parts of London includes an internal lead-lined gutter running inside the roof space from front to rear draining the front parapet gutter in order to obviate any need for a down pipe at the front. A blockage to the rear outlet to one of these results in all the front roof slope rainwater overflowing above the top floor ceiling with disastrous consequences.

Inter-war

There was very little new building during either the First World War or the Second World War. Most builders were called up to fight. Construction began in about 1919 and ended in 1939 so during this period we had the growth of the inter-war suburbs.

Depending upon the part of the country, walls were either solid or cavity brick. If cavity watch out for signs of rusting wall ties. Slate dpc and suspended timber ground floors are common. Watch out for high ground levels bridging dpc and blocked sub-floor ventilation. Generally very little trouble is experienced with buildings from this period provided that they have been well maintained during their lifetime.

Nineteen-fifties and -sixties

A system of government licences was in force until 1953 and a licence was required for any private house building so as to divert maximum resources into the huge local authority building programme. Private house building generally dates from 1953 onward, typically simple solid construction. Because timber was imported (much from Canada) and Britain was short of dollars every effort was made to minimise the use of timber. Ground floors were normally solid concrete. Minimal architraves and skirting boards were used with simple flush-faced doors filled with stramit. Watch out for perished felt dpc, settlement of floor slabs and inadequate damp-proof membranes in solid floors. Also leaking Finlock gutters and brittle sarking felt which tears easily. Exterior painted softwood all suffered badly from wet-rot attack.

During the 1960s we found an increasing use of flat roofs surfaced in mineral felt, often on poor-quality decking of stramit or chipboard with no vapour barrier and poor insulation.

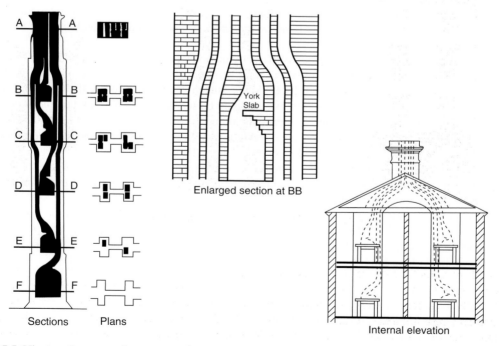

Sections   Plans   Enlarged section at BB   Internal elevation

Fig. 15.5 Nineteenth century flue construction. During inspection it is wise to count the chimney pots and trace the flues within the building. The bringing together of flues to the chimney is called 'gathering'.

Nineteen-seventies, -eighties and -nineties

Design and construction improved gradually over this period. Thermal insulation increased. The need to encourage the development of more 'brownfield' and poor quality sites led to some foundation problems, however, and the problems associated with the action of tree roots in clay soils were highlighted in a drought in 1976 which led to a general tightening up of site investigation procedures and foundation specification with an increasing use of bored piles on shrinkable clay sites.

Local characteristics

When carrying out a Building Survey on an older building one should beware of any structure which appears to be out of keeping with its locality. So if we find a building constructed of materials which are not indigenous to its location we should be looking for defects. Buildings using traditional materials and construction methods may be the result of 300 or 400 years of trial and error.

London

In London we generally find most of the construction to be in brick.

Few residential or small-scale buildings dating from before the Great Fire (1666) remain. The city and the east was developed first so that the older buildings here are generally Georgian or later with the attractive residential squares following the uniform pattern of the Rate Houses. The main problems are at the bottom in the basements and cellars which are not tanked in the modern manner and which invariably suffer dampness, and at roof level due to leaks at parapets. Structural movement in the form of leaning and bulging walls is common with lots of tie bars used to restrain outwardly bulging brickwork, especially to the tall flank walls at the end of terraces. Except at basement level the internal walls are invariably stud partitions clad in willow laths and lime plaster, with soft plaster often held together by the lining papers. Extensive use was made of bonding timbers and other woodwork built into the walls and prone to rot and beetle attack.

The west was developed later with much of Kensington and Chelsea very late Georgian or Victorian. These developments followed the pattern of the Rate Houses with whole areas laid out to a grid as in Pimlico, built by Thomas Cubit around 1860. Much stucco rendering used with increasing use of Italianate decoration in stucco from 1860 to 1900 as may be seen in South Kensington and Bayswater. Most of what looks like stonework is in fact stucco including the ornamental columns and pilasters to the entrances. When the frost gets behind the stucco and it falls off the columns are found to be of brickwork, often using broken brick and snapped bricks as an infilling.

The later Victorian buildings tend to be better constructed than the Georgian with better foundations and thicker walls, suffering less from structural movement but still prone to the usual problems induced by dampness at basement and roof level.

Many of the basements have been converted into flats. Sometimes also the vaults at the front extending even below the pavement. I hardly ever find one of these, that is, totally free from damp. In general if you want a dry basement you have to construct a swimming pool inside out. Anything less than this will not work in the long run.

The regions

In addition to local brick we find a great deal of construction in stone. We also find a great variety of roof coverings including not only local slate and tile but also materials such as stone slabs and shingles.

In relation to stonework some notes on the weather qualities of the commoner building stones may be useful.

Building stones

Bath stone.   A mined stone, fairly soft and easy to work and responds well to cleaning. Varies in colour from cream to light cream with a more open texture than Portland stone. Poor weathering qualities in industrial atmospheres and attacked by acid rain.

Beer stone.   Fine-grained white limestone from east Devon. Poor weathering qualities but commonly found in many churches in the west of England and, in particular, Exeter Cathedral. The quarries had to be specially re-opened to find matching replacement stone for Exeter Cathedral when badly weathered stonework needed matching replacement.

Clipsham stone.   Coarse-grained light brown limestone from the Rutland area used for restoration work on the Houses of Parliament and for general building in the Midlands and Oxford area.

Darley Dale.   Derbyshire sandstone, fine-grained and light grey to buff colour. Good weathering qualities but difficult to clean.

Doulting.   Soft, coarse-grained limestone from Somerset, coloured grey to light brown, with a crystalline structure containing shell fragments. Very poor weathering especially in industrial atmospheres and generally unsuitable for exposed parapets and cornice work.

Forest of Dean.   Gloucestershire sandstone, varying in colour from blue to grey to green, often used for paving, monumental work and exposed facings as it weathers very well and is fine grained.

Ham Hill.   A soft Somerset limestone containing shells, rich yellow-brown colour. Cleans well but unsuitable for city atmospheres.

Hopton Wood.   Derbyshire limestone, cream colour and containing fossils. Can be polished to give a marble-like appearance, however, it has poor weathering qualities in city atmospheres.

Kentish Rag.   Commonly used in London area, a grey-white limestone which cleans well and is often used for rubble walling.

Portland stone.   White stone commonly used in towns and cities for classical detailing. Quarried in Dorset. Probably the best limestone for city atmospheres.

Wootton stone.   Red sandstone from Cheshire and south-west Lancashire, sometimes called Runcorn stone. Durable and fine-grained but best used inland as it weathers badly in marine atmospheres.

York stone.   Very durable grey-brown Yorkshire sandstone, fine-grained and very hard. Commonly used for paving, sills and external dressings to buildings. Difficult to clean.

Granite.   Hard stone varying in colour from pink to grey to brown. Considerable variation in quality depending upon source. The best comes from Aberdeen or Dartmoor (Devon). Ideal for most applications including engineering works. Beware of soft brown granite from west Cornwall, however, as some of this has poor weathering qualities and is in the first stages of conversion into china clay.

Slate.   Thin beddings of slates used for wall construction must be laid with the courses sloping downwards to the outside to prevent damp penetration. Much used in the Lake District, Isle of Man, Wales, central Cornwall and borders of Dartmoor.

Mortars

Modern Portland cement mortars are generally unsuitable for the purpose of repairing or repointing older brickwork and stonework. Generally the mortars used should be weaker, both in terms of strength and permeability, compared to the bricks and stones they bond. Lime mortar should therefore generally be used for any repairs to older-wall construction. If this is not done then the trapped moisture within the walls will be unable to escape through the mortar and will be absorbed into the stone or brick. This will result in efflorescence and frost damage to the surfaces causing spalling and lamination of the brick and stone facing.

Pitched roof coverings

Prior to 1666 the London roofs of ordinary buildings were typically thatch, with predictable consequences during the Great Fire. Subsequently the London roofs were mostly covered with slates or plain
clay tiles. The tiles were hand made and held on the battens using small timber pegs. Rot to the pegs was
common and slipped tiles were often held in place using a lime mortar torching applied from beneath
within the roof void. Slates were fixed using small iron nails, not galvanised and prone to rust away completely after 150 to 200 years.

In the regions we find a much greater variety of pitched roof coverings manufactured from local materials or excavated from local quarries. In assessing these old roofs some local knowledge is essential.

Slates vary in quality and lifespan. The better type of Welsh slate can often be re-used more than once
and an old slate roof suffering from nail sickness can often be re-roofed using the original slates on new
battens and sarking felt with modern galvanised steel nails and then give another 100 years of service.
Remember that sarking felt should be laid fairly loose rather than tight on the rafters so that there is a
slight dip between rafters – any water running down the felt is then less likely to come into contact with
the battens. Additional ventilation is also needed when a roof previously not underfelted is lined with
sarking felt.

Plain tiles, hand made from local clays, also vary in quality. When these are laminating, with the nibs
soft and broken, then they are unsuitable for re-use and a new tile covering is needed. A number of manufacturers now produce machine-made tiles which replicate the texture and appearance of old hand-made
tiles and are a good substitute for re-roofing purposes.

Interlocking clay tiles will also be found, such as the Bridgewater type in the West Country and
Belgian types imported from Belgium and Holland and used principally in the South East and Midlands.
Unlike plain tiles (which have a double lap) these only overlap for a short distance and rely on the roof
pitch and tile profile to prevent driving rain and snow getting behind. As originally laid they would have
no underfelting so damp penetration is common and if a tile slips the rain will pour in. Matching replacement tiles are often difficult to find – modern concrete interlocking tiles have different profiles. So an old
roof with one of these coverings will generally need re-roofing. Remember that the roof timbers may have
been subjected to damp conditions over many years so rot and beetle attack is likely to battens and rafters.

Stone slabs are used to cover roofs in some areas, notably the Cotswolds where the stones, with their
mossy and weathered surfaces, are an attractive feature of the typical old village houses. Beware of any
old roof of this type which shows evidence of repairs in Portland cement mortars or coatings of bitumen
or other material intended to improve water tightness and extend its life. Any old stone roof which has
been subjected to repairs of this type is almost certainly on its last legs with significant problems to timbers and other supporting construction beneath. Clients should be advised to budget for a new roof and
if the property is in a Conservation Area they may be required to use matching stonework from an
approved local quarry.

Buildings of special interest

There is a system of control in force in Britain under the Town and Country Planning Acts and the
Government compiles a list of buildings which are of special architectural or historical interest. Owners of
such buildings have to be notified before listing and the listed buildings are recorded in the local land
charges register held by the local authority. So the solicitor acting for a purchaser will normally discover,
when making a search of the register, if a property is listed as being of special architectural or historical value.

In addition to the normal planning rules an additional permission – 'listed building consent' – is then
required before alterations can be made which would affect its character as a building of special architectural or historical interest including the demolition of any part of the building.

In addition to the procedures for listing specific buildings there is also provision under the Town and Country Planning Acts for localities to be protected by being designated Conservation Areas. Here again the fact that a building lies within a conservation area will show up on the local search made by the purchaser's solicitor.

Demolition within a conservation area requires specific permission and development within a conservation area has to be subject to prior public notice. Development within a conservation area will normally only be permitted within the parameters of specific guidance for the area issued by the planning authority.

When carrying out a survey of an older building which is listed or in a conservation area the surveyor should be particularly aware of two possible implications for the client.

First, the client may have proposals to alter and extend the property. Clients with such proposals must be advised to consult the planning authority and if the purchase is dependent upon the works being approved it would generally be advisable to make the applications and obtain the necessary approvals before legal commitment.

Secondly, repairs may be required and the cost of repairs may be considerably increased if there are special rules regarding the materials which can be used. For example, roof coverings may have to be in local stone or slate, or hand-made tiles, rather than cheaper modern alternatives. Ornamental plasterwork and stucco may have to be replicated. Original fenestration details may have to be retained necessitating the use of specialist joinery contractors. Copper and lead coverings may have to be used for roofs rather than cheaper modern substitutes.

There is always a problem finding good builders – especially in London – and a particular problem if specialist trades are required with the necessary skill and experience of working on listed buildings.

Grants and penalties

Outstanding historical buildings may receive repair grants. This generally applies only to Grade I listed properties. The Grade II listed buildings (overwhelmingly the majority) will not normally be eligible for grants. The local planning authority can serve notice on the owner of a listed building if it is in disrepair, requiring repairs to be carried out and in extreme cases of neglect they have powers of compulsory purchase if deliberate dereliction has been allowed by the owner. The planning authority may themselves carry out repairs and then charge the cost to the owner.

So the purchase of a listed building in poor repair may be something of a gamble for the client. He or she is unlikely to be lucky enough to obtain a grant to help with repair costs and the repairs themselves are likely to be more costly with budget over-runs almost inevitable.

16
Reports on leasehold properties

> The managing director read the lease, and could not understand it. Then he read the accompanying solicitor's letter and could not understand that either. So he asked for a letter from them clarifying their first letter and could not understand that . . .
>
> *Extract from client's letter of instruction*

If a client is contemplating taking either a new lease or an assignment of the existing lease of a property, certain special considerations apply to the advice to be given, taking into consideration the particular position of the lessee (the tenant) under the law of dilapidations.

Leases of houses

A number of different possibilities may arise. In Manchester, Cardiff and certain other localities, dwelling-houses have been sold traditionally on a leasehold basis, generally for a 99- or 999-year term. The reasons for this are essentially historical and arose because the landholdings in these areas were in the hands of certain families. Initially, the idea was good since after 99 years and the expiration of the leases, the general redevelopment of an area might have been contemplated and the ground landlords would have retained some control over the properties for the benefit of all concerned. Unfortunately, the results in practice were that leasehold dwelling-houses tended to fall into an increasing state of disrepair as leases began to run out because the lessees were unable or unwilling to improve them.

As a result of pressures for reform of the leasehold system, the 1967 Leasehold Reform Act was passed which gave leaseholders with certain types of long leases of houses (but not flats) the right to buy the freehold, or the right to a specially-extended term at a moderate ground rent subject to proving the required period of owner-occupation and certain other conditions.

Unless the lease term is short (say less than 30 years) it is felt that it would be unusual in practice for the ground landlord of a leasehold house to take very much interest in its condition, and it would be most unlikely that a schedule of dilapidations would be served except in unusual circumstances. In most cases the report on such a leasehold house will follow the same lines as would be the case with a freehold property.

Perhaps the main point which could be made to a client contemplating buying a leasehold house would be to suggest that he might acquire the freehold as well. He could do so by asking the vendor to serve the necessary Notice of Leaseholder's Claim and transfer the property with the benefit of that Notice. If this is not done, the purchaser cannot take advantage of the Act himself until he has first served out the necessary term of occupation. The client should be advised to consult a solicitor on this point. In the long run, a freehold house is far better investment.

Where the lease of a house is running out and has only a short time to remain, then the necessity for a schedule of dilapidations may well arise. Shorter leases of houses will be encountered in some fashionable London districts, and rights to enfranchise under the Leasehold Reform Act may be different or may not exist, depending upon the size and value of the property concerned. A careful reading of the lease in such circumstances is recommended before giving the appropriate advice (see Chapter 20).

Leases of flats

Leases of flats and apartments are the rule rather than the exception in England and Wales for the reasons discussed in Chapter 14. An almost infinite variety of forms of such leases will be encountered in practice. However, the surveyor will find that these generally fall into one of two basic types: either there will be a lateral division of responsibility within the structure of the building with each leaseholder responsible for his or her own part of the structure and having no liability for other parts (mostly encountered in cases of small two- or three-storey buildings in flats, either purpose built or converted), or the lessee will be responsible for the interior of his or her own unit and share responsibility for the exterior and structure as a whole with other flat owners. In the latter case, arrangements for the proper maintenance and upkeep of the exterior and structure may simply be on an *ad hoc* basis as the need arises. Alternatively, this may be the responsibility of the freeholder who would then recover costs from the lessees, or commonly, may be undertaken by a management company or residents' association.

It will be seen in such cases that a lessee may have a potential proportional liability for parts of the building, or even parts of other buildings on the estate some distance away from his or her own property. Particularly onerous repair liabilities may arise in the case of flats in older mansion blocks in the inner cities.

Common practice is for such a mansion block to be acquired by a speculator, and for individual flats to be sold off on new long leases with a ground rent and service charge for the maintenance and repair of the exterior, roof and common parts. Initially, in order to sell the leases, the freeholder will wish to keep the service charges low, perhaps artificially low. Once the units are all sold, however, there is no such need and when the true state of repair of the structure begins to become known, large increases in the service or supplementary charges may be levied to cover the cost of repairs which should have been dealt with at the outset during refurbishment and prior to sale.

Before undertaking an inspection of a flat in such a mansion block, the surveyor should be quite sure that the survey will be sufficient to gauge whether or not any exceptional service charge increases may be due. If this is to be done, it will be clear that the inspection cannot be confined merely to one specific flat. The significance of this is discussed more fully in Chapter 14.

In all cases of surveys of leasehold residential property, the Royal Institution of Chartered Surveyors (RICS) Code of Practice advises that the surveyor should read the lease before undertaking his inspection. It is advisable to establish the extent and nature of the repairing liabilities under that lease, the client's potential responsibility for executing repairs and also his liability to pay for repairs executed by others. If the lease is not available, the surveyor must set out clearly the limitations.

If no copy of lease is available, the client's solicitor should be approached and asked to confirm the nature of the repairing covenant which may be assumed. The surveyor may then, on receipt of such instructions in writing from the client's solicitor, refer to this in compiling the report.

The Code of Practice suggests that when taking instructions, in addition to the procedures recommended for freehold property, the surveyor should insist either on having a copy of the lease or on having advice on the nature of repairing liabilities. He or she should then confirm instructions in the manner recommended for freehold property, indicating also the extent to which he or she proposes to inspect other parts of the property or estate in order to advise on unacceptable or unusually heavy repairing liabilities outside the demised premises.

Leases of commercial premises

Dealing with leases of commercial premises, the recommendations for procedure could be similar to those for leasehold residential property in that it is highly desirable for the surveyor to read the lease before undertaking the inspection and to have it available for references as when compiling the report. The report should cover:

(a) the repairs or other works which are required for the purposes of the client's business occupation and necessary to comply with the various local and national regulations, and
(b) the future repairs or other works which may be required in order to comply with the repairing covenants of the lease.

In the case of (a) above it is necessary to ascertain the nature of the client's business, how he proposes to use the premises and the staff that will be employed. In respect of (b), the surveyor will have regard to what the lease says and the current relevance of the law of dilapidations (see Chapter 20).

Section 18(1) of the Landlord and Tenant Act 1927 sets out certain principles in the assessment of damages for breaches of repairing covenants. It provides that the measure of damage for breach of a covenant to keep premises in repair, or leave premises in good repair when quitting, cannot exceed the diminution in the value of the landlord's reversion occasioned by these breaches. A limit is therefore imposed on the amount of damages which may be recovered by a landlord. These are assessed by valuing the premises as they are and as they would be if the covenants had been complied with, the measure of damage being the difference between these two figures. The cost of undertaking remedial works may be the same thing as the difference in value, on the assumption that a purchaser of the lessor's interest would reduce any price offered by an amount equal to the cost of remedying those defects. The cost of undertaking works may, however, be an amount quite different from the measure of damages in some cases. It is possible for there to be considerable breaches of repairing covenants by a lessee and for the landlord to have no right to compensation at all (e.g. if the landlord is proposing to redevelop the property in such a way that condition is immaterial to market value).

The Courts have held that the cost of undertaking works is *prima facie* evidence of the diminution in the value of the reversion, in which assumption can be rebutted by evidence that damage to the lessor's interests may be better assessed by some other means, such as 'before' and 'after' valuations. In *Drummond* v. *S. & U. Stores Ltd.* (1980) 258 EG 1293 (*Estates Gazette*, 27 June 1981, p. 1293) it was held that in the absence of clear evidence as to the damage to the reversion, the cost of undertaking necessary works was the broad measure of the diminution of value and the landlords were awarded the bulk of the cost of repairs. The landlords in this case were also awarded 3 months' loss of rent as part of the damages, being the estimated time reasonably necessary to leave the premises empty while the works were done, and also, as the landlord in that case was not registered for value added tax (VAT), that VAT on the cost of repairs would also be recoverable.

The surveyor advising a client on a commercial lease should have uppermost in mind the possibility of either an interim or final Schedule of Dilapidations being served on the client and should warn the client about all matters which could give rise to the service of such a schedule or a claim for diminution of value. Substantial sums may often be involved in such cases. One means of protecting a client at the commencement of a lease is to arrange for the lease to state that the lessee will not have to hand the demised premises back on expiration of the term in any better condition than they were in at the commencement as evidenced by a Schedule of Condition. An agreed Schedule of Condition is then prepared by the lessee's and lessee's surveyors, signed and attached to both lease and counterpart lease.

During the course of taking instructions to inspect a leasehold property, the surveyor should ascertain whether or not a Schedule of Condition exists which indicates the condition of the premises as at the beginning of the term. If such a Schedule exists, then it would be advantageous for the surveyor to have a copy and to take this along when carrying out the inspection. The condition of the property as described in the Schedule of Condition may then be compared with the condition found on inspection.

If a building is to be inspected when a client proposes to take a new lease, and if that building is old or in poor repair, then the surveyor may recommend that a Schedule of Condition be prepared and agreed with the prospective lessor prior to exchange of the lease and counterpart. This may be especially important where a modern form of full repairing lease is being granted on a building which is old and which may suffer from signs of structural distress, past foundation movements, faulty roof construction and other matters potentially expensive to rectify. In general, the lessee's liability is for repair, and where a defect can be repaired then the liability is not to renew or rebuild, nor will there be any liability to make a building better than it was previously. In practice, however, there is considerable room for argument as to the extent of repairs which may be required.

Perhaps a roof covering is in poor condition and should be stripped off and renewed if it is to be made satisfactory. However, a repair in the form of overhaul and patching of the covering may suffice in the short term. There could be considerable room for disagreement between lessor and lessee as to the extent and nature of repairs in such a case. A Schedule of Condition detailing the condition of the roof covering at the commencement of the term would provide a standard against which any repairs or renewals may be judged adequate or otherwise.

It may be that a property suffers from dampness. Generally, there is no liability for the lessee to provide damp-proof courses or other works where none previously existed, although in such circumstances he will be liable to deal with the effects of dampness if these have caused deterioration of the interior. If damp-proof courses were in good condition at the commencement of the lease term but failed and became porous during that term, then a liability to renew damp-proof courses could arise. In such a situation it may be difficult to confirm all the facts and if a Schedule of Condition is discovered which clearly shows that the property had a damp problem when the lease began, then this could be of considerable benefit to the lessee's case. Conversely, if a Schedule of Condition indicates that the property was dry and the damp-proof courses were in good condition at the commencement of the term, then this would assist the lessor in requiring remedial works to be undertaken at the lessee's expense, or in obtaining damages for diminution of value.

On occasion, a building may suffer from structural movements of one kind or another. In old buildings such matters rarely arise overnight. More often than not they are part of a continuing process with crack damage, bulging and leaning walls and other manifestations having appeared very gradually over a long period of time. Rarely could a lessee be found liable to undertake a structural repair such as underpinning, but there could arise a liability to provide buttresses to leaning walls or tie bars to stiffen the structure. An original Schedule of Condition is helpful in such cases if it accurately describes the structural condition of the building in detail as existing at the commencement of the lease term.

In addition to the repairing covenants in a lease there will also be other covenants, some not directly relevant to the survey but others which may affect the inspection and preparation of the report.

There will generally be a 'user clause' in the lease indicating uses which may be unacceptable and those which may be permitted. For example, a client who wishes to use premises for purposes other than those acceptable in the lease must first obtain the lessor's consent for the change of use. If the lease provides that such consent may not be unreasonably withheld, then the lessor, in determining whether or not to grant consent, is subject to a test of reasonableness which may be challenged. If in the lease there is no provision as to reasonableness the lessor may or may not grant consent for a change of use at his or her discretion. Clearly, prior to any exchange of contract to acquire a lease, a client must obtain a determination in such cases.

There may be objections to a specific retailing use which conflicts with other retailing uses in adjoining premises which are owned by the same lessor. A lessor who owns a number of retail premises in the same shopping district will wish to ensure a balance of different types of retailing in the area for two reasons, one being the need to present a broad range of retailing outlets to make the centre attractive and profitable, and the other being to limit competition between shop units selling identical products or services which could otherwise make one, or both, competing units unprofitable.

Objections to use may arise out of the nature of the uses themselves since dust, noise, smells or other problems could arise to the annoyance of the lessor or other adjoining lessees. Certain specific trades will often be particularly referred to in a lease as being unacceptable. These will often include panel-beating, fish-and-chip shops, heavy industry in a light industrial area, or trades such as glue factories or fertiliser storage which could cause offence.

There will generally be some mention of 'assignment' in the lease. A lease with an absolute bar on assignment cannot be sold and is of value only to the lessee in possession as it has no value as a mortgage security. Where assignment is permitted with consent, the law holds that such consent may not be unreasonably withheld, even if the lease omits to state this, so that assignment to a lessee of comparable standing will normally be permitted subject to the lessor providing a Licence to Assign, generally on payment of a small fee to a solicitor.

Covenants in the lease barring structural alterations to the buildings demised are generally found. Such covenants will generally refer to the need for specific permission to be given by the lessor for any alterations to the premises, but an absolute bar on alterations may be provided as an alternative.

A lessee who wishes to undertake alterations to business premises which will constitute improvements and which may increase the value of the premises may, if he or she wishes, serve notices under the 1927 Landlord and Tenant Act. This is often overlooked, with the consequence that a lessee loses any rights to compensation for improvements or additions to the landlord's property when he or she vacates. Advice about 1927 Landlord and Tenant Act procedures needs to be given at the outset before improvements are made and should be dealt with at the planning stage.

Changes in leasehold law

New legislation is now in place to allow the owners of flats and maisonettes to purchase their homes with commonhold title whilst at the same time allowing enforceable arrangements to arise to deal with the maintenance and upkeep of common parts and the payment for shared services. Such reforms have been long overdue and pressure is now growing from the owners of leases granted some years ago where the shortening lease term is having a serious effect on the resale value of the property.

Commonhold title is currently available only for purchasers of new flats and the take up of this new system of land tenure has been very slow with no current proposals to extend it to cover existing leasehold property.

Consideration is also being given to the problems associated with the management of blocks of flats, which is a matter inevitably tied up with the existing leasehold arrangements which have been found to be unsatisfactory in many cases. Again reform of the law to provide access to special courts to settle disputes and order works to be carried out may be the answer as serious difficulties are arising at the present time with the management of larger blocks of flats, especially in the cities where many such buildings are old and in poor repair.

Existing arrangements permitting leaseholders of flats to club together and buy the freehold compulsorily, or extend their leases, have been found to be complex and difficult to apply in practice.

There is also a continuing problem with valuations when leaseholders wish to extend the lease or buy the freehold, arising out of the natural desire of the freeholders to share in the marriage value that this creates, so that the freeholders often make a large capital gain over and above the investment value of their interest if they are allowed a proportion of the marriage value.

17

Reports for prospective mortgagees

If language is not correct, then what is said is not what is meant; if what is said is not what is meant, then what ought to be done remains undone.

Confucius

Mortgagees are primarily concerned with ensuring that the security they take for the purpose of the mortgage will be readily re-saleable at a price which will allow for the repayment of any monies due and cover the incidental costs involved, including the costs of sale.

Institutional mortgagees will also wish to ensure that their actions are seen to be in the public interest and they will want to foster the goodwill of their mortgagors and consumers generally. With this in mind, the main building societies and banks now release copies of their surveyors' reports to the mortgagors.

It should be mentioned at this stage that the mortgagor grants the legal mortgage in return for the loan and is the borrower (and purchaser). The mortgagee takes the legal mortgage, grants the loan and is the lender. It is not uncommon for the terms to be wrongly transposed and it is important to use the correct nomenclature.

Reports for prospective mortgagees may take the form of pro formas or be the traditional typed reports on the surveyor's own paper. Such reports may or may not include a full building survey but they will inevitably include some comments on the condition of the structure since it is not possible to value a property without first forming an assessment of its general condition.

If the property is in poor repair the surveyor will be expected to give the mortgagees some guidance on the significance of this and the conditions which should be applied to any loan. Such conditions may take the form of retentions from the advance money pending completion of essential works and a further surveyor's inspection, or may simply require an undertaking from the mortgagor to complete the works within a specific period of time.

The advice to be given on repairs required will be dependent to some extent upon the amount of advance involved. Generally, low-percentage advances involve low risk levels for the mortgagees and in many cases the requirement for repairs will not be critical. In such circumstances, a written undertaking from a mortgagor of good repute to complete the works within a specified time would seem adequate.

High-percentage advances involve higher risk levels to the mortgagees, and experience tends to indicate that the risk of default by a mortgagor tends to be higher in the earlier years of the mortgage term when the outstanding balance of capital is at its highest. In such circumstances, it may be necessary for the mortgagees to retain part of the agreed advance and to release this against certification from the surveyor when essential repairs have been completed. Building Societies are especially vulnerable in this respect because they provide mortgage advances to house buyers in circumstances where the percentage advances are high (sometimes as much as 100%) and the security is often an older building in poor repair.

The mortgage transaction is achieved by means of a mortgage deed in which the mortgagor agrees to terms controlling the repayment of capital and interest and may also agree to certain requirements for the maintenance and repair of the property used as security over the period of the loan, also the insurance of the building against specified perils. The mortgagor retains the right to recover the property title free from the encumbrance of the mortgage, on payment of the amount due plus costs and penalties if the loan is redeemed early. The right of the mortgagor to redeem the loan is known as his 'equity of redemption'.

Virtually all mortgages the surveyor has to deal with will be mortgages granted by a charge on the property expressed by way of a legal mortgage. Since 1925, this has been the principal method of lending money to a property owner against the property security and is the only legal form of mortgage apart from a much less usual procedure for the grant of a lease to the mortgagee with provision for cessation on redemption. Occasionally, an equitable mortgage is encountered which may be a contract to enter into a legal mortgage if required, or a verbal agreement accompanied by the deposit of the title deeds or lease of the property with the mortgagee. This latter procedure was occasionally adopted by bank managers when granting small loans to achieve security without the need for a full legal mortgage. It does not confer a legal interest in the property and is not suitable for larger-percentage long-term advances, such advances invariably taking the form of a legal charge.

The security for the loan lies in the ability of the mortgagee to take possession of the property and sell it to recover the loan together with incidental costs. The Council of Mortgage Lenders (CML) has issued a Statement of Practice, 'Handling of Arrears and Possessions', which sets out recommended good practice in this respect. No mortgagee likes to take possession and institutional mortgagees such as banks and building societies will go to considerable lengths to avoid such procedures. They may give the borrower an extended time to pay or a capital repayment 'holiday' to see him through a difficult time. Only as a last resort will they take steps to obtain possession of the property, especially in the case of residential owner-occupied homes, although the number of homes taken into possession has increased in recent years. Local authorities tend to be lenders of last resort in some sections of the residential property market and the incidence of default by local authority borrowers is rather higher, as is the incidence of default by borrowers on the security of property investments and commercial premises.

Generally, the procedure adopted when all else fails is for the mortgagee to apply to the Courts to take possession. A mortgagee in possession will normally be able to require the borrower and his family to vacate if the property is an owner-occupied house and will also be able to obtain vacant possession in other cases where the borrower, or borrowing company, is also the occupier. The property would then be sold in the open market for the best price obtainable at the time. Any surplus left over after the loan, incidental expenses and accumulated interest have been paid must be paid to the mortgagor. If the property is subject to tenancies and the tenants are protected by virtue of the Rent Acts, Landlord and Tenant Acts or Agricultural Holdings Acts, then the mortgagee takes legal possession subject to the subsisting tenancies. The existence of tenancies may therefore substantially reduce the re-sale value of the mortgage security. When a surveyor inspects the security he or she should look out for any signs of tenancies about which the mortgagee may be unaware and suggest, if appropriate, that the existence or otherwise of tenancies be confirmed.

An alternative procedure would be for the mortgagee to apply to the Courts for a foreclosure order which would have the effect of extinguishing the mortgagor's equity of redemption and transferring the property to the mortgagee. For technical reasons this procedure is less frequently used than the procedure for the mortgagee entering into possession, but in certain situations foreclosure can have advantages.

Another alternative procedure is for the mortgagee to take possession but with a view to managing the property, collecting the income and paying any outgoings. The net income could then be used to reduce the mortgage debt and pay the interest charges with any arrears. In modern times, such a procedure would not normally be used for vacant property unless the premises concerned proved difficult to sell in the open market but were potentially lettable. These circumstances could arise with certain types of commercial or industrial premises, premises affected by planning blight or having other problems which would be a disadvantage to the potential purchaser but not to a potential tenant.

It may be that a mortgagee in possession will take over premises which are tenanted, either because the premises were already let and the loan was granted on the security as an investment, or because the tenancies have been created after the grant of the loan. Generally the mortgage deed will provide that the mortgagee's consent is required for any letting of the property which secures the loan, but lettings may be undertaken unlawfully without the mortgagee's consent. In such situations it is more likely that the mortgagee in possession will not seek to sell the property immediately since the value may be very low either because of the existence of tenancies or the climate in the property investment market may be unfavourable. It may then be advisable for the mortgagee in possession to manage the property and use the income to pay off the debt as well as meeting the interest due.

In valuing a property for mortgage purposes the surveyor will then need to consider two matters. First, the re-sale value of the property should be sufficient to cover the amount of the loan plus incidental costs. If this is not so, then the mortgagee will suffer a loss. In certain circumstances where the loss suffered is due to the surveyor's valuation being too high, the surveyor may be held to be negligent. Secondly, in the case of premises which are tenanted, or are to be tenanted, the surveyor may need to consider whether or not the net rental income from the property will be adequate to cover the repayments. Clearly it would be unwise in most cases for a mortgagee to advance money on the security of a property investment in circumstances where the income from the investment is less than the mortgage repayments.

Certain types of property represent good security for mortgage advances. Owner-occupied houses with vacant possession and in good condition are good security since there is a broad market for such houses in most localities in all save the most direeconomic circumstances. Residential investments are a less satisfactory security especially if the net income is insufficient to cover the repayments on a substantial percentage advance. In modern times, much of the value of tenanted investments lies in their potential for future vacant possession and such values are prone to the effects of general economic circumstances, interest rate changes and changes in legislation.

Commercial, industrial and other types of building will have market values and potential rental incomes which will vary greatly in different localities and in different economic circumstances. The more specialised or remote such a property is, the greater will be the degree of risk of a fall in value. Some types of older industrial building may become unsaleable or unlettable in time if a suitable alternative use cannot be found. The implications of such a situation for the mortgagee will clearly be very serious.

In making a valuation of a property for mortgage purposes, the surveyor should follow the ordinary principles of good practice. Since this book is not intended to cover the principles of valuation, it is not proposed to discuss techniques. However, there are certain matters which will have to be confirmed during the course of the survey of a building for a prospective mortgagee because the information is needed in order to prepare a valuation.

The first of these matters to be ascertained is the condition of the building. Although the survey may not be a full building survey and the inspection may be abbreviated, nevertheless it is not felt that any building can be valued unless the surveyor has inspected it inside and out, however superficially, and formed some view as to its condition. Clearly the possibility of serious defects in a building, especially an older building, has to be considered and if such defects are present they will need to be taken into account in formulating a valuation. If such defects are suspected but cannot be confirmed in the time available or without special exposure, then a provisional or conditional valuation may have to be issued and the prospective mortgagee will then have to consider whether to pursue the matter any further in the light of the percentage loan applied for.

It is usual for the valuation report in many cases to contain caveats or conditions having regard to the possibility of concealed structural defects, high-alumina cement concrete in widely-spanned commercial structures, and similar hidden problems. In many such cases the mortgagee will require the mortgagor to obtain a technical assessment of possible structural problems of this type as a pre-condition of any advance. This is a useful way of safeguarding the mortgagee's and mortgagor's positions since any such matters which come to light will be of interest to them and may be crucial.

Measured surveys

The second matter to be assessed is the physical size and shape of the building so that it may fairly be compared with other similar buildings for valuation purposes. This will generally require a measured survey to confirm the floor areas, clear spans, ceiling heights and usable frontages. The Royal Institution of Chartered Surveyors publishes a Code of Measuring Practice, which is periodically updated and all surveyors should have a copy. It is recommended that the practices set out in this Code be adopted wherever possible. The Code recommends that gross internal area (GIA) as defined should be used for industrial and warehouse space, although agency practice in many areas is to use gross external area (GEA) when dealing with new industrial and warehousing buildings. The Code also recommends measuring procedures for shops and residential buildings.

The purpose of an accurate measured survey is to enable the surveyor to retain in the records details of properties valued so that details of similar properties may be compared on a like-for-like basis. As long as the basis of measurement is the same, the basis of comparison will be valid and the statistical records kept for valuation purposes will be correct. With industrial and warehouse space, I would commend the use of GIA as being the fairest means of comparison, especially if older properties are being compared with more modern buildings. The usable floor space in an older building with thick walls and other obstructions may be proportionally less than the usable space in a modern structure with clear spans and thin walls. A comparison of GEA in such circumstances may be misleading.

In addition to the two foregoing important matters which have to be resolved when surveying a building for a prospective mortgagee, there are also a number of other incidentals arising during the survey which may be significant. The surveyor may see evidence of tenancies, rights of way, shared access or other complications which could be significant to the mortgagee and which should be mentioned in the report. The mortgagee may also require a pro forma report to be completed, in which case the surveyor will need to check the questions in that report and confirm that they can be answered fully from the notes taken on return to the office.

The mortgagee may require a plan to be prepared showing the curtilage of the property, main plot dimensions and other means of identification such as the distance from the nearest road junction and the north point. The purpose of an identification plan is to compare the property inspected with the title plan to ensure, among other things, that the surveyor has actually inspected the correct building (the possibility of the wrong house being inspected is not as far-fetched as it may seem, especially on a new estate) and that his valuation does not include any property or land which is not part of the holding.

Having completed the survey of the building for the prospective mortgagee, the surveyor will then have to compile the report. The report may be a full building survey, or something rather less, on the surveyor's own paper; alternatively it may be completed using a pro forma report form. The surveyor should not normally advise the lender in relation to the percentage advance or the length of the mortgage term, these being matters for the mortgage underwriter, but surveyors are often asked to advise on general lending policy.

General lending policy

General lending policy is concerned with the nature of the properties upon which the prospective mortgagee may feel able to lend and the duration of the loans; the terms controlling the frequency of capital repayment and the rate of interest; and the percentage advance considered suitable in the circumstances.

A trustee, clearing bank or building society lending at relatively low interest rates over long periods will expect near-absolute security for loans. They will, quite reasonably, expect that in the event of a mortgagor's default, they will suffer no loss and the proceeds of sale will cover not only the original advance but also some margin for accumulated interest and costs. Traditionally, an advance of two-thirds or three-quarters of valuation would normally be the maximum loan in respect of good, readily saleable security. Something rather less than this, perhaps 50% or less, may be considered suitable in respect of poorer security. The margin is intended to cover not only accumulated interest and costs in such circumstances, but also the possible vagaries of the market for certain types of property. If the market for a property is doubtful then it may be that no advance at all can be contemplated by an institution at low interest rates and the prospective mortgagor may have to produce some other security. Alternatively, he or she may approach a lender prepared to accept a greater risk which will undoubtedly involve a higher interest rate on the loan.

During the post-war period, the building society movement has tended to advance rather higher percentages on owner-occupied dwelling-houses with loans of up to 90% or 100%. To enable them to do this, they will normally require some additional collateral such as a mortgage guarantee from central or local government, from an insurance company or a charge on another property or on a life insurance policy.

A surveyor may be called upon to advise a prospective mortgagee on loans being made at higher than normal interest rates on the security of properties which may be declined for this purpose by a mainstream mortgage lender. In this situation it is as well that the lending policy of the clients be clarified at the outset and that the surveyor is able to exercise judgment in the light of that policy so as on the one hand to protect the client's interests and on the other hand to allow sufficient flexibility for a suitable agreement to be reached.

Mortgagees who advance money at higher than normal interest rates will normally have to accept that the quality of the security offered or the status of the borrower is lower than may be ideal. The higher interest rates which they may charge are intended to reflect this additional risk. Such lenders may be prepared to advance money on the security of residential investment property where the net income is lower than the mortgage repayments, relying upon the borrower's ability to finance the balance from some other source. They may be prepared to lend money on the security of vacant sites, or sites with buildings which are destined for redevelopment where there may be little or no income generated from the property, relying upon the borrower's ability to redeem the advance when redevelopment has been undertaken. They may be prepared to advance money on the security of unusual buildings which are in secondary or remote locations or buildings suffering from defects of one kind or another.

In carrying out a survey for such a mortgagee the surveyor cannot necessarily apply all the principles of lending policy which would be applicable to those mortgagees who charge more conventional lower interest rates. Each situation has to be considered on its individual merits, having regard to the fact that if all principles of good lending policy were to be applied then the institution concerned would be unable to compete with other institutions offering terms at lower rates. On the other hand, if an imprudent lending policy is adopted, the end result may be a spate of defaults should there be a downturn in the property market.

Local authority mortgage

Local authorities make advances on the security of residential property and, in most cases, the intention of providing this service is to enable house-purchase to be contemplated by sections of the community who would otherwise be unable to become owner-occupiers. Local authority mortgagors tend to be younger, seeking to buy houses in the cheaper price ranges with fairly large percentage loans.

Whereas a bank or building society may limit advances to 75% or 80%, the local authority may lend up to 100% of valuation, and in some neighbourhoods, act as lenders of last resort. Many surveys of properties for local authorities are carried out by surveyors who are employed for the purpose by the authority as staff surveyors. Occasionally the local authority will employ an outside surveying practice to do this work.

A feature of surveys for local authority mortgages is that the property being inspected is often older and in poor condition. In the larger conurbations, the building may comprise a converted flat or apartment. In some localities the local authority may be the only potential lender in the market so that the saleability of older houses, converted flats and apartments in some districts will depend almost exclusively upon the funds which the local authority has available at the time. Since the money for local authority lending is subject to the effects of national economic policy, there will be times when loans are granted fairly freely and other times when funds are 'tight' and the availability of council mortgages may dry up altogether.

In carrying out a survey and preparing a report on an older house in poor condition, or a converted property, the surveyor should consider whether it would be reasonable to impose conditions which could make the property more saleable in the future by making it potentially mortgageable to a lender other than a local authority. Converted flats, for example, may lack adequate soundproofing or fire-resistance between units but if these matters are dealt with, perhaps at fairly modest expense, the converted unit may be more mortgageable and therefore more saleable in the future, especially in times when council loans are difficult to obtain. At the time of writing very little local authority lending is being undertaken and their role in this respect has been largely taken over by the mainstream mortgage lenders.

Leasehold interests

The property being offered as security for an advance may be leasehold. Residential property held on a long lease will not normally be suitable for a long-term advance by way of mortgage if the unexpired term is 30 years or less. If a term of more than 30 years is available, the repayments of capital should be so made that all capital outstanding is progressively reduced until the mortgage is redeemed when the lease still has at least 30 years to run. The reason for this is obvious – a leasehold interest is a wasting asset which will fall in real value during its life. The effects of monetary inflation will mask the fall in value for a time but once the end of the lease is in sight, the fall will be rapid. When the lease expires, the value for mortgage purposes will be nil.

If an advance is made on the security of a short leasehold interest in a residential property, the risk to the mortgagee is high because in the event of default by the mortgagor, there may be considerable delay caused by the legal procedures for obtaining possession. While this delay occurs, the value of the security may be falling. With freehold security, the mortgagee will at least normally know that increasing property values and inflation may be on his side should he find that delay is incurred in proceeding with a remedy for default.

With commercial and industrial premises, a mortgage advance is often made on the security of a lease where the period available before the lease expires, or the remaining period before the next rent review, is quite short. Such leases may have only a very speculative value. In times of economic boom it may be that substantial premiums will be obtainable for such leases, partly reflecting the profit rent (that is the difference between the rent payable and the open market rent which would be obtainable) and partly elements of goodwill, tenants' improvements, fixtures and fittings and temporary scarcity. In times of economic depression, no premium may be obtainable and tenants may be glad to able to vacate by making a straight assignment of the lease without premium. While a mortgage advance may be secured by a legal charge on the lease in such situations, the real security for the advance will not be the value of the lease but will be the ability of the borrowers to repay. If the borrowers are in difficulties, perhaps during a period of economic depression, the fact that a mortgage is secured by the value of the lease may be worth nothing to the mortgagee.

A highly cautious lending policy is therefore recommended where a short leasehold interest is involved. The surveyor should ensure that all the lease covenants have been complied with, especially covenants on usage and repair and should carefully read the lease to confirm whether or not there are any unusual covenants which could limit the re-sale value of that lease; tight restrictions on uses, in particular, may make a lease unsaleable and therefore potentially unmortgageable. If the building is in poor order, and especially if the condition is such that a Schedule of Dilapidations may be served, the property may need to be put into good repair as a condition of any mortgage advance, and probably as a precondition of some, or all, of the advance being released.

When surveying a leasehold property for a prospective mortgagee it is suggested that the surveyor should make a point of reading the lease. The only exception might be in the case of residential property where normal lease terms may be assumed for the purpose of the mortgage valuation and where the mortgagee's solicitors will be able to confirm this and refer back to the surveyor if unusual terms are found to apply. In many cases a building society or bank valuation will have to be undertaken on this basis and a provisional valuation given without the surveyor having seen the lease.

Insurance

A report for a prospective mortgagee will often include advice as to a suitable amount of insurance cover and the mortgage deed will normally provide that the building is to be insured in the joint names of the mortgagee and the mortgagor.

In addition to insurance against fire, there is a wide range of other perils against which cover may be recommended, including impact damage, explosion, storm, tempest, flood, burst pipes and tanks, subsidence, landslip, ground heave, public liability, plate glass, trade fixtures and fittings and loss of rental income or loss of profits. Not all such perils will arise in every case. Subsidence, landslip and ground heave cover is normally available for residential buildings but not in all other cases. Plate glass and consequential loss cover will apply to certain types of commercial premises.

The insurance should cover all buildings used as security for the advance unless they are not significant to the re-sale value. The sum insured should be adequate to cover the cost of reinstatement in the event of a total loss. If the sum insured is excessive, the mortgagor will be penalised by having to pay excessive premiums. If the sum insured is insufficient then difficulty could arise in the event of a claim not only for a total loss, which rarely occurs, but also for a partial loss. The insurers may, in a case of obvious under-insurance, average the claim by paying only in proportion to amount of cover as it relates to the full reinstatement value. For example, if a building is insured for £100 000 and its true rebuilding cost would be £200 000, the insurers may be prepared to pay only 50% of any claims made for insured perils.

In the case of building societies they will generally act as agents for the insurance company who insure all their mortgage securities, often under a single block policy. The surveyor undertaking an inspection of a building for a building society or other institutional lender may need to provide an estimate of the reinstatement cost for insurance purposes and also advise the lenders of any obvious insurance risks which they should then relay to the insurance company.

In low-lying or coastal areas, property could be subject to flooding, and often enquiries have to be made where a building is low-lying in order to ascertain whether or not there is a flood risk. In such cases the insurers may refuse to provide the cover for flood damage or may require additional premiums. Properties with thatched roofs or other types of potentially flammable construction may be subject to premium loading. Buildings located at potentially hazardous points adjoining road junctions or fast sections of highway may be prone to impact damage. Certain types of construction could be prone to storm damage, especially lightweight structures in exposed locations. Subsidence, landslip or ground heave cover may be restricted where a building already shows signs of existing or past structural distress.

It is a rule of insurance law that the parties must disclose any matters which could be relevant to the degree of risk and it is incumbent upon a surveyor who notes any such matter to bring it to the attention of the mortgagee in the report so that the insurers may be suitably advised.

The amount of insurance cover assessed by reference to rebuilding costs current at the time should be periodically reviewed and it is good practice to provide that the sum insured be altered annually in line with changes in building costs. Most insurers of buildings have some procedure for automatic adjustment of the sum insured along these lines. If no such procedure applies, it is suggested that the mortgagees should themselves undertake to review the amount of insurance cover each year to safeguard their position.

The Building Cost Information Service of the Royal Institutional of Chartered Surveyors publishes an annual *Guide to House Rebuilding Costs for Insurance Valuation* on behalf of the British Insurance Association. This guide refers to costs of demolishing and clearing away an existing structure and rebuilding it to its original design in modern materials and with modern techniques to the same standard as the original property but to comply with Building Regulations and other statutory requirements. This information is essential where reinstatement costing advice is to be given.

Second and subsequent mortgages

In additional to the provision of first mortgages, it is possible for financial institutions or individuals to lend on the security of second mortgages, or even third or subsequent mortgages.

Clearly, the degree of security for second or subsequent mortgages will be less than that afforded by first mortgages. The second or subsequent mortgagee will be able to recover his amount owing only after the amounts owing to the first mortgagee have been satisfied as far as this is possible out of the proceeds of sale. In effect, the mortgagees have to take their turn in the queue if a mortgagor defaults and the property used as security is sold. It may be that a second or subsequent mortgagee will find it difficult or impossible to recover his debt if the amount owing under the first mortgage is substantial and if there are accumulated costs and interest charges also owing to the first mortgagee.

Having regard to the higher risks attaching to second or subsequent mortgage advances, the interest rates charged are generally higher and the terms for capital repayment often of shorter duration. If a surveyor is asked to survey a building to be used as security for a second or subsequent advance it is a good idea to find out the amounts of capital outstanding under any first or former mortgages and to confirm the amount of the borrower's equity in the property and the extent to which any further mortgage advance will be covered.

If the degree of cover for the second or subsequent mortgage is poor then great caution should be exercised in preparing any valuation and in making any recommendations, bearing in mind the particular need to safeguard the future re-sale value of the security in the light of the state of repair of the premises and all other circumstances.

Mortgagees in possession

Occasionally a surveyor has to return to re-inspect property used as security in circumstances where the borrowers have defaulted, perhaps to advise on the appropriate sale price to quote or to prepare a claim for dilapidations. It is a good idea for surveyors to inspect such property once in a while to remind them of the difficulties which often face mortgagees in such situations. Rarely will the property be in good order; often it will be in very poor condition with serious neglect and perhaps even deliberate damage. The legal process can take considerable time and a borrower in financial difficulties will certainly not be able to afford a high standard of maintenance. In many cases the borrowers may have removed fixtures and fittings, perhaps even stripped the property of anything of value.

The author recalls inspecting a house in North London on one occasion following repossession by a building society to find that the borrowers had removed a large number of internal fixtures including all the interior doors, the bath panel and the lavatory seat. On another occasion the occupiers of a small light industrial workshop had stripped out all the electrical wiring and copper plumbing before making their escape.

In law anything permanently affixed to the land becomes part of the security but defaulting mortgagees do not always respect the legal niceties. The mortgage deed may specify that interior and exterior decorations are to be dealt with at specified intervals but few borrowers in financial difficulties will be able to comply.

The mortgagee in possession has an obligation to safeguard the value of the security and obtain the best price for it when selling bearing in mind that the mortgagor may have a financial interest in the proceeds of sale, and the mortgagee has to mitigate its loss. The CML publishes guidance, *Handling of Arrears and Possessions, Statement of Practice*, setting out recommended good practice.

18

A typical Building Survey Report

For if the trumpet give an uncertain sound, who shall prepare himself for the battle? So likewise ye, except ye utter by the tongue words easy to be understood, how shall it be known what is spoken?

St Paul

This is a traditional narrative style Building Survey Report prepared on a house in the London area for a prospective purchaser. Although this was an actual job the address and some of the details have been changed to avoid identification. The report follows the general advice given in this book; matters of description are given in the present tense whilst findings are given in the past tense and the wording follows as closely as possible the usual advice that the surveyor should (1) describe what was seen and not seen, and the tests made, then (2) give some analysis of the result in terms of the condition of the property and finally (3) tell the client what – if anything – he or she needs to do, including where necessary arranging further investigations.

Many of the sentences and paragraphs will use standard text with the intention of streamlining the draughting process, but surveyors should always bear in mind the need to modify and adapt text to suit the property, and the client, if the end result is to read well.

TAPP AND POKE
Chartered Surveyors
Crumbling Chambers
Weathering
London SW3

1st May 2006

A. Client Esq
100 Concerned Street
Worrying
Hants.

Dear Mr Client,

1, SUBURBAN ROAD, LONDON, SW15

1.0   Scope of Instructions

We have carried out a Building Survey of the under-mentioned property in connection with your proposed purchase. The report has been prepared in accordance with the signed and agreed terms and conditions of engagement.

Fig. 18.1 Suburban Road.

2.0   Client's Name

Mr A. Client

3.0   Address of Property Inspected

1, Suburban Road, London SW15.

4.0   Date of Inspection

25th April 2006

5.0   Introduction and General Remarks

This report should be construed as a comment on the overall condition of the property and is not an inventory of every single defect, some of which would not significantly affect the value of the property. For further details please refer to the conditions of engagement.

At the time of inspection the weather was cold and dry. The house was occupied and furnished with floors covered. Insulation material in the loft has been disturbed only to the extent necessary to identify the nature of the construction beneath and we have not removed loft insulation material generally. There was a large quantity of storage and personal effects in the loft which limited our access to some areas. Fitted floor coverings have been lifted in sample areas only in order to identify the nature of the construction beneath and we have not taken up fitted floor coverings generally. No exposure of hidden parts of the structure has been carried out.

6.0   Description and Age of Property

A two storey terraced Edwardian family house dating from about 1910 and located in an established residential road of similar properties, convenient for all amenities. More recently some modernization and refurbishment of the house has been carried out including modern double-glazed windows at the rear and a replacement roof covering. The house retains many of the original period features and is generally presented well for sale.

The house is located on the south-west side of the road facing north-east so the morning sun will shine towards the front coming around to the rear in the afternoon. The prevailing weather will tend to drive towards the back of the house from the south-west.

There is a small rectangular shaped garden at front and rear with a pedestrian footpath at the end of the garden which we assume to be a joint private right of way and your legal adviser should confirm the details of this by reference to the title plan. There is no garage or on-site parking. Street parking in this area is controlled by resident's permit and meter.

The site where the house stands is generally level.

7.0   Accommodation

First floor – landing, three bedrooms, bathroom, separate water closet (WC).
Ground floor – porch, hall, three living rooms, kitchen.
Outside – front and rear garden.

8.0   Current Use

Residential dwelling.

9.0   Roads and Access

The highway appears to be adopted and your legal adviser should confirm this. The road in front is a cul-de-sac but this is not very well designed since there is no hammerhead for turning at the end it may become rather congested at busy times. Please note our previous comments regarding the shared footpath that gives access to the end of the garden.

10.0   Tenure

We understand that the house is being sold with an unencumbered absolute freehold title and with vacant possession and your legal adviser should confirm this.

You should ask your legal adviser to confirm who is liable for the various boundary fences by reference to the title plan. Some of the property titles in this area contain restrictive covenants which require an original freeholder's consent for alterations or extensions and you should ask your legal adviser to confirm whether any restrictive covenants apply in this case.

11.0 State of Repair: External

11.1 *Chimney Stacks and Flashings*

There are two chimney stacks constructed in brickwork with lead flashings, one located over the left-hand side party wall at the front with four open pots and another located over the right-hand side party wall above the kitchen with three capped pots. We inspected the chimney stacks externally where possible from ground level and also from the skylight.

No works were seem to be necessary to the chimney stacks at the time of inspection but normal ongoing maintenance in the form of occasional pointing repairs are likely to be required from time to time in the future. It was noted, however, that the lead flashing around the base of the chimney stack at the rear is lifted and displaced where this is dressed down over the tiles and this needs to be fixed back in to maintain weathertightness at this point. This is a fairly minor repair which can be dealt with quite easily by a roofing contractor.

11.2 *Roofs*

The roof has a modern covering of interlocking concrete titles on battens and underfelting, supported by a framework of sawn softwood timbers. In the main front part the roof timbers comprise 100 mm × 50 mm rafters and joists with two 200 mm × 50 mm purlins, two 100 mm × 50 mm struts, one 100 mm × 50 mm hanger and two 100 mm × 50 mm binders. This is a normal layout of roof timbers for a house of this type and age. Over the front bay there is a projecting gable-end to the roof finished with rough cast brickwork and the roof slopes to this are formed with 100 mm × 50 mm rafters and joists with lead linings to the valleys on either side. At the back of the house the roof extends rearwards over the rear wing with 100 mm × 50 mm rafters and joists used.

We examined the roof slopes where possible from ground level externally and from within the roof space. No worked were seem to be necessary to the various roof slopes at the time of our inspection.

The party walls on either side of the house are carried up in the form of parapet firewalls at roof levels. These being in brickwork with lead flashings finished on top with tile copings. Parapet firewalls were inspected from ground level during our survey and also to some extent from the skylight over the rear wing. No works were seen to be necessary to this part of the construction at the time of inspection where seen, but normal ongoing maintenance in the form of occasional pointing repairs will be needed in order to maintain appearance and weathertightness.

11.3 *Roof Voids*

The roof space is reached via a trapdoor set into the landing ceiling. The trapdoor itself takes the form of an old coloured glass window which has been adapted for the purpose and above this there is a modern single-glazed plastic skylight set in to the roof slope to provide natural lighting on to the landing. Some care needs to be exercised in using the loft hatch because the old glass panel is potentially quite fragile.

We inspected within the roof void where possible but our inspection was limited to some extent by insulation material and a large amount of storage. No works were seen to be necessary within this area at the time of inspection, but it should be noted that only about 25 mm of fibreglass quilt insulation is laid over the ceilings, which is a very low level of insulation by modern standards and we would recommend the provision of at least 100 mm of fibreglass quilt tucked down neatly between the ceiling joists with ventilation provided to the space above and up to 250 mm could be laid for maximum benefit. Ventilation into the roof space should be provided in the form of air-vents set into the roof slopes with small holes cut through the sarking felt beneath, and we would recommend that you arrange for a roofing contractor to add two air-vents to each of the roof slopes and then increase the depth of loft insulation to a better modern standard.

11.4   *Parapets, Parapet Gutters and Valley Gutters*

Please note our previous comments regarding the parapet firewalls on either side of the roof. Valley gutters are provided on either side of the front gable and also where the two roof voids join between the main front part and the rear wing. In each case lead linings are used and no works were seen to be necessary to these at the time of inspection.

11.5   *Gutters, Downpipes and Gullies*

Unplasticised polyvinyl chloride (PVCu) gutters and rainwater pipes are provided. Periodic checks during heavy rainfall would be advised since the sponge seals to this type of guttering often work loose. This house has solid walls (in contrast to modern cavity construction), so any leakage of water down the walls externally can penetrate directly to the inside causing damp. These walls probably incorporate a certain amount of timber built into the construction in the form of lintels or wall plates and these timbers will be vulnerable to rot or other defects if they are allowed to become wet.

Wisteria has been growing up the back wall of the house and over roof and this should be removed and cut back so that it does not enter and block the gutters. There is a poor detail to the gutter at the back of the house where the roof water from the main roof drains down into the gutter over the rear wing. During heavy rainfall this water has been overflowing down the wall causing gutter splash in the corner resulting in some green moss growth on the brickwork and this could penetrate through to the inside causing dampness at skirting level in the adjacent living rooms. Some modification and improvement of the gutter layout is needed to prevent gutter splash outside in this corner. A roofing contractor should be able to deal with this fairly easily using leadwork.

Stormwater is normally taken to a combined foul water and stormwater sewer in this part of London. At the back of the house the rainwater runs into a combined stormwater and foul water gully and thence into the sewer manhole in the garden. There is also a stormwater gully at the front of the house but we do not know where the water from this then runs to. It may be that it runs into the drain in the road. The gully at the front of the house should be periodically checked to ensure that water is running away freely without leakage.

11.6   *Main Walls*

We have not undertaken any trial bores in order to confirm the nature of the subsoil underneath this house however, the Geological Survey Map for the area indicates that the subsoil is likely to be Kempton Park Gravel which is a river terrace material found along the Thames floodplain. These subsoils are normally found to provide a satisfactory foundation base for a simple low rise structure of this type however, it is important to ensure that the drains around the property are maintained in good order because leaking drains washing away the fine particles inside the subsoil can cause damaging subsoil erosion and foundation movements.

We have not dug any inspection pits in order to examine the foundations but normally these houses were built with conventional shallow strip foundations consisting of a concrete strip with brick footings laid between 600 mm and 900 mm below ground level. This is a fairly shallow foundation by modern standards.

The house is built with 225 mm solid brick walls around the perimeter and the party walls are also of solid brickwork wherever seen. The internal wall between the front and rear part of the house is a solid brick wall but elsewhere the internal walls are timber stud partitions clad in lath and plaster.

We examined the walls internally and externally where possible. No evidence of unusual movements or settlements to the main walls was found at the time of inspection to the areas seen.

We did find evidence of some downward settlement of the partition walls around the bathroom at first floor level where the bathroom construction rests on the floor over the rear living room. This has resulted in some distortion to the door openings to the rear bedroom and WC with a noticeable slope to the top of both door heads. Such matters are very common in Edwardian construction of this type where partition walls are supported on the floors and where the floors also take the weight of bathroom sanitaryware and other point loads. There is no evidence of recent cracking or significant movement at this point and the settlement noted is not related to the foundations.

No specific remedial action is considered to be necessary in relation to the distorted door openings at first floor level but you could, if you wish, have these door openings straightened up by adjusting the doorframes and rehanging the doors prior to the next redecoration.

11.7   *Damp-Proof Course*

Slate damp-proof courses are used wherever seen at the base of the main walls, mostly covered by a cement rendered plinth externally, and the plaster and skirtings internally.

As a general rule the outside paving should be no higher than 150 mm (two brick courses) below the damp-proof course or internal floor level, whichever is the lower, with a slope away from the walls. The ground clearance at the front of the house meets this standard but at the rear there is some bridging of the damp-proof course by soil and paving immediately behind the kitchen and we recommend that the flower bed in the corner and the paving itself be relaid so that there is a stepdown of at least 150 mm from the kitchen floor into the garden, with a slope away to a suitable stormwater drain.

11.8   *Sub-Floor Ventilation*

The ground floors are mostly of suspended timber construction except in the porch and kitchen where there are areas of solid concrete floor. These timber ground floors require ventilation and air-vents should be provided at a maximum of 1.5 m centres around the perimetre with an air-vent in each corner in order to maintain a flow of air under the floor to keep timbers dry. One of the air-vents to the front bay has a hole in it which will permit the access of mice or other vermin and we recommend that this be provided with a new grille.

11.9   *Windows*

The windows at first floor level are single-glazed sliding sash type and appear to be original. These windows are in poor condition with some loose joinery and wet rot to the corners. If you buy the house and if you propose to keep the windows we recommend that you employ the services of a specialist box sash window repair company to carry out a full overhaul of the windows including renewal of the lower woodwork where needed, followed by redecoration. If you are in any doubt regarding the potential costs involved in repairing and overhauling the first floor windows a quotation should be obtained prior to legal commitment to purchase.

At ground floor level there is one set of single-glazed painted softwood casement doors from the centre living room to the garden and these will require ongoing maintenance attention and decoration. The other ground floor windows are PVCu double-glazed type.

In relation to the double glazing a problem often arises due to condensation forming between the glass and when this happens the only remedy is to reglaze the windows which can be expensive. There was no evidence of condensation between the glass here at the time of our inspection, but condensation problems in double glazing are very common and tend to come and go, to some extent, depending upon the temperature of the glass and other factors.

We would therefore recommend that enquiries be made with the vendor to ascertain when the double glazing was installed, who did it and whether it is covered by a long term transferable guarantee, back by insurance.

If the double glazing was installed after April 2002 the work should have been undertaken by a FENestration Self-Assessment scheme (FENSA) approved contractor, or it should have received specific consent from the local authority under Building Regulations, and your legal adviser should confirm whether either circumstance is applicable in this case.

11.10 *External Joinery (to include doors and door frames)*

A normal level of ongoing maintenance and periodic redecoration will be required paying particular attention to the fascia boards behind the gutters and the window woodwork as previously described.

11.11 *External Decorations*

These will require attention at the normal intervals.

11.12 *Other Relevant Matters*

There is a slight lean to the rear end of the left-hand side party firewall at the back of the rood and the condition of the brickwork to this firewall should be kept under observation. If any further leaning or cracking appears it may be necessary to take down some of the brickwork here and rebuild it, a matter which would have to be agreed with the owner of the adjoining property. No specific action is considered to be necessary yet on current evidence.

There is also an area of open jointed pointing to the arch over the rear bedroom window and the arch brickwork here will need to be repointed in due course in order to maintain appearance and weathertightness. At the front of the house there are a few small areas of soft or open pointing, most noticeably to the outer corners of the front bay. These will require maintenance attention in the form of simple pointing repairs in due course.

The front gable-end wall is surfaced in painted rough cast and this is uneven and may be hollow in places under the existing paintwork so when the outside of the house is next decorated you should allow a contingency provision in your budget in case the gable-end rendering needs to be partially or totally resurfaced before new decorations are applied.

12.0 State of Repair: Internal

12.1 *Ceilings*

The life of lath and plaster ceilings will depend to some extent on the quality of the original job and the degree of exposure to dampness and vibration. These ceilings are now approaching one hundred years of age and may be close to the end of their economic life so when rooms are next decorated a point should be made of carefully probing and checking the ceiling plaster, which should be replaced or underlined with modern plasterboard before new decorations are applied. Similar considerations apply to the lath and plaster stud partitions. It is not always necessary to take down the old lath and plaster ceilings in these circumstances, since plasterboard can be applied from underneath using longer nails and then reskimmed. If you are in any doubt regarding the potential costs involved in dealing with ceiling plasterwork in the future, a quotation and report from a plastering contractor should be obtained, prior to legal commitment to purchase.

12.2 *Internal Walls and Partitions*

Normal ongoing maintenance in the form of occasional plaster repairs and periodic redecoration will be required with a need for general resurfacing in plasterboard eventually when the lath and plaster becomes excessively loose and off-key. Please note our previous comments regarding settlement of the partition walls around the bathroom which has resulted in distorted door openings to the WC and rear bedroom.

12.3   *Fireplaces, Flues and Chimney Breasts*

If you propose to use flues for open fires or appliances, they must first be swept, the condition of the linings confirmed and a smoke test carried out. In a house of this age built in a soft lime mortar, it is likely that the flues would need to be relined before they can be safely used, and a reputable fireplace contractor should be consulted. The front living room fireplace is currently used for a gas 'coal' fire, whilst the rear living room fireplace has an open hearth which is apparently used for coal or log fires. The flue to the dining room has been sealed off, with no air-vent inserted, and there was probably an old kitchen range or similar located here. In the bedrooms there are three cast iron fireplace surrounded with open flues but these should be regarded as ornamental only and unsuitable for use until the flues have been tested.

Where flues are redundant they should be provided with air-vents at the base and capped pots at the top to keep them dry. The three pots over the rear wing have been capped but the four pots to the main chimney stack are open and caps should be fitted to the redundant flues here to prevent direct weather penetration downwards.

12.4   *Floors*

At first floor level the floors consist of butt jointed boarding resting on 175 mm × 50 mm joists where seen. At ground floor level the floors are mostly of butt jointed boarding resting on 100 mm × 50 mm joists, supported by sleeper walls where seen. We have not lifted fitted floor coverings in most areas due to the consequential damage which would result.

In the kitchen, the floor is of solid concrete construction on the left-hand side and some form of timber floor is provided on the right-hand side, both areas being covered by vinyl flooring stuck down with no access to confirm what lies beneath.

We recommend that some further investigation be carried out to confirm the nature of the floor construction in the kitchen, and at this stage you should allow a contingency provision in your budget in case some modifications are needed here. Generally, a solid concrete floor should incorporate a damp-proof membrane, contiguous with the damp-proof courses in the surrounding walls in order to keep the floor surface dry. If there is a timber floor this would normally have an air space beneath which requires ventilation although no evidence of ventilation was apparent when we inspected and there is a flower bed at the back outside which is bridging the damp-proof course as previously noted.

In these circumstances if there is a timber floor which is unventilated, there is possibility of rot or other defects to the timbers. It is possible to have a timber floor resting on battens over a damp-proof membrane in circumstances where ventilation is not needed, but investigation would be required in order to confirm if this circumstance applies in this case.

12.5   *Internal Joinery (to include internal doors, staircases and built-in fitments)*

Standard quality architraves, skirtings, panel doors and other interior woodwork is provided with a basic range of cupboards and worktops in the kitchen. You will no doubt wish to examine the various interior fixtures and fittings yourself in order to confirm whether they meet your requirements and you should allow in your budget for any improvements needed to suit your own tastes.

12.6   *Internal Decorations*

The house is generally presented well for sale although there are some marks to the decorations in places and some torn wallpaper on the landing, probably as a consequence of the activities of the residents' children.

12.7   *Dampness*

Tests for dampness were made on a cold dry day with the house ventilated and heated. The results obtained in different circumstances could be different. No unusual dampness was found during our inspection. Low-level moisture readings were obtained in places in the kitchen and to the front bay, but on current evidence we would attribute these to condensation.

Condensation can be a problem in old houses with solid walls, especially when double glazing is installed which makes the structure effectively airtight and it is important to maintain the correct balance of background heating and ventilation. The temptation to dry laundry, etc. on the radiators should be avoided and some form of extraction ventilation should be used from the bathroom and kitchen, notwith-standing that these rooms may have windows. We would recommend a ducted cooker hood rather than a simple filter hood in the kitchen and an extractor vent operating automatically from the light switch with a time delay in the bathroom.

At present in the kitchen there is a filter hood and above this an air-vent set into the wall so much of the steam passing through the filter hood is likely to remain within the kitchen and this has caused some condensation staining to the decorations over the cooking area. A larger cooker hood ducted directly to the outside through the wall would be a recommended improvement.

12.8   *Woodworm, Dry Rot and Timber Infestation*

Woodworm and other timber defects are commonly found in houses of this age in this part of London. Local treatments and repairs to timbers may be required from time to time. This should be regarded as normal ongoing maintenance. Timbers are less likely to suffer such defects if kept dry in a centrally heated house and if the sub-floor ventilation remains unobstructed.

We found scattered flight holes of the common furniture beetle (woodworm) to the timbers in the understairs cupboard and there has probably been the normal history of furniture beetle activity in this property in the past.

No new flight holes or bore dust were found in the areas seen.

We recommend that enquiries be made with the vendor to ascertain if the timbers in the house have been treated for Furniture Beetle activity and if it is covered by a long term transferable guarantee issued by a reputable specialist contractor, backed by insurance.

If there is no evidence of successful past timber treatment then you should expect to find that there will be some further evidence of furniture beetle activity and other defects in the floors and elsewhere, the extent of which can only be confirmed by exposure and at this stage, pending confirmation of the position, you would be advised to allow contingency provision in your budget in case further treatments or repairs are required.

The best current advice is that the excessive use of chemicals for timber treatments in houses should be avoided and treatment should be targeted to areas of obvious activity where new flight holes and bore dust are apparent. The wholesale precautionary treatment of timbers in houses should be avoided.

12.9   *Thermal/Acoustic Insulation*

A normal level of acoustic insulation to the party walls on either side may be assumed, so some noises could be heard from the neighbours from time to time, this being normal with terraced housing.

As regards thermal insulation, the solid external walls will lose heat more rapidly than modern cavity construction and you are likely to find the single glazed windows cold and draughty. The main windows at ground floor level have been double glazed. Please note our previous comments regarding the gener-ally low level of loft insulation over the ceilings which could usefully be increased to reduce heat losses through the roof. It would also be a good idea to double glaze the loft hatch to reduce heat losses through this part of the property.

12.10   *Other Relevant Matters*

There is a small dent to the laminated wood floor in the dining room over a knot hole or similar in the boarding beneath.

13.0   Services

13.1   *Electricity*

A meter with residual current circuit breaker (RCCB) unit and circuit breakers is located under the stairs. We recommend that the electrical installation be inspected and tested by an National Inspection Council for Electrical Installation Contracting (NICEIC) registered electrician, tests to include insulation, polarity and earth continuity, with a check to ensure that all plumbing and gas services are bonded to earth. Any works necessary to comply with British Standard 7671 should be undertaken.

13.2   *Gas*

The gas meter is located in an external metre box. We recommend that the gas service, together with any gas appliances included in the sale, be inspected and tested by a Council of Registered Gas Installers (CORGI) registered contractor, paying particular attention to boiler flue and ventilation requirements and the gas fire in the front living room.

13.3   *Water, Plumbing and Sanitary Fittings*

The company stopcock is located in the pavement at the front. We have not seen the underground main supply pipe but if this is an original lead pipe it will now be due for replacement in modern polypropylene and this will involve digging up the front pathway. We recommend that enquiries be made with the vendor to ascertain what is known of the age and type of main water supply pipe. If it turns out that this is lead replacement is advised and until such time as it is replaced, it would be advisable to run the kitchen cold tap for a short time in the morning before filling kettles, etc., so that water is not being drunk that has stood in a lead pipe overnight. This advice applies particularly if there are small children in the house.

The remainder of the plumbing is in copper pipework where seen with outlets to bath, basin, WC and sink. There are no storage tanks, all services being directly off the mains. Sanitaryware and fittings are all of a fairly poor basic quality. The bath enamel is quite badly chipped in places and the popup waste to the basin is broken. We would anticipate that most potential purchasers would probably allow for refitting the bathroom and WC to a better modern standard in due course.

13.4   *Hot Water and Heating Installation*

There is an ocean combination boiler with balanced flue located on the wall in the kitchen and this heats pressed steel panel radiators and provides demand-activated hot water. We recommend that the system be inspected and tested by a CORGI registered heating engineer and it should be serviced.

13.5   *Drainage*

There is a shared drain running underneath the rear garden from left to right with one manhole enclosing a branch connection to an open gully and cast iron soil vent pipe. Open gullies are no longer recommended for hygiene reasons and we would advise having this replaced with a modern back-inlet gully where the waste pipes discharge below the grating. Waste water is taken into the gully via plastic hopper head and downpipe from the bathroom and plastic waste pipes. The roof water from the back of the house also runs into this gully.

The drain flow was satisfactory at the time of our inspection. We must emphasise that we have not tested the drains under pressure and we have not inspected inside manholes in any other adjoining gardens.

A shared drain installed before the 1st October 1937 (which applies in this case) is normally a public sewer, notwithstanding that it may pass through private gardens so responsibility for this drain probably rests with the main drainage authority (Thames Water). You should ask your legal adviser to confirm the status of the drain. If follows that if any problems are encountered with the drainage system in the future or it needs unblocking the main drainage authority should be approached in the first instance.

14.0   The Site

14.1   *Garages and Outbuildings*

None.

14.2   *Grounds and Boundaries*

The garden areas are surrounded by timber fences with a low wall to the pavement at the front. There is an original Edwardian black and white tiled path leading up to the front entrance door.

There is a tree growing in the neighbour's garden, beyond the footpath at the rear, being a sycamore approximately 8 m high and approximately 8 m from this house. There is also a small pavement tree in front of the adjoining property to the east which appears to be an ornamental maple or similar. Tree roots can be damaging to structures and services below ground, however, no evidence of any damage having been caused by these trees was noted at the time of our inspection. The trees will require some ongoing maintenance and management but this would not be a matter within your control if you purchased this property. The sycamore tree is likely to cause some overshadowing of the garden at the rear.

14.3   *Flooding*

The house is located on the Thames floodplain, close to the river. We are not aware of any history of flooding having affected this property or those nearby but your legal adviser should make the usual environmental searches and enquiries. We do not know if insurance companies load premiums or impose special conditions in respect of houses in this neighbourhood and your legal adviser should confirm the position.

14.4   *Environmental and Mining Matters*

We are not aware of any environmental or mining matters relevant to this survey, but your legal adviser should make the usual enquiries. *Reference to the Radon Atlas of England and Wales*, published by the National Radiological Protection Board indicates that the level of radon gas rising from the ground in this locality is very low or insignificant.

14.5   *Building Regulations, Town Planning and Other Relevant Statutory Matters*

There would not appear to have been any alterations to this house which would have required planning permission.

Some alterations may have been made which required approval under Building Regulations including the replacement roof covering, however, in practice many small builders do not bother to apply for this sort of work.

14.6   *Contamination and Hazardous Materials*

No evidence of contamination or hazardous materials was found during our inspection to the areas seen.

15.0   Summary and Recommendations

Where we have recommended further investigations or reports from specialists in this report, these should be obtained together with estimates for the necessary work, prior to legal commitment to purchase.

In the various foregoing pages of this report we have mentioned various items for repair that are considered necessary and advise that, prior to commitment to purchase, you obtain estimates from local building contractors for the works as required.

We have not carried out tests to ascertain whether any deleterious or hazardous materials or techniques have been used in the construction of this building, or has been incorporated subsequently. Neither have we conducted tests to ascertain if the land is contaminated.

For the purpose of this report we have assumed (unless stated to the contrary) that there are no easements, covenants, restrictions or other outgoings of an onerous or unusual nature that would materially affect the value of the property.

The report provided will be confidential to the named client(s) for the specific purpose to which it refers. It may be disclosed to the client's professional advisers. It shall not be disclosed to any other person nor reproduced in full or in part without prior consent.

Your legal advisers should confirm that the property is insured from the moment of exchange of contracts for a sufficient sum against all usual perils including fire, impact, explosion, storm, tempest, flood, burst pipes and tanks, subsidence, landslip and ground heave.

The house has generally been well maintained and is presented well for sale. The main matters arising out of our inspection are:

1. The original lath and plaster ceilings and stud partitions.
2. The poor condition of the original window frames.
3. Improvements advised to loft insulation.
4. Some lowering of ground levels around the property.
5. Some investigation advised to confirm details of the kitchen floor construction.
6. Routine tests of services advised.
7. Replacement of the main water supply pipe if this is original lead.
9. Some minor works to roof flashings and gutters.
10. Past history of woodworm activity.

16.0   Assumptions and Provisos

These are as specified within the agreed conditions of engagement, a copy of which are attached, unless otherwise stated below.

This report has been electronically approved by:

A. TAPP FRICS

Fig. 18.2 Testing No. 1 Suburban Road for damp. The green light on the left indicates that the wall is dry. If the light moves up into the red sector it would indicate increasing levels of moisture in the wall plaster. This would not necessarily indicate rising damp due to failure of damp-proof courses however; there are many other reasons why wall plaster could be damp.

Fig. 18.3 Sloping door head due to downward settlement of partition walls.

Fig. 18.4 Loose and lifting lead flashings will allow driving rain to penetrate.

Fig. 18.5 Significant expenditure on repairs or replacements of window frames will be needed prior to the next redecoration.

Fig. 18.6 Green algae and moss growing at the base of a wall due to gutter splash.

Fig. 18.7 Just the right size hole for mice.

Fig. 18.8 One hundred years old drains. Drain flow satisfactory.

19
Legal considerations

The buyer is not entitled to remedy the defect and charge the cost to the surveyor. He is only entitled to damages for breach of contract or for negligence . . . So you have to take the difference in valuation . . . In other words, how much more did he pay for the house by reason of the negligent report than he would have paid had it been a good report?

Lord Denning MR – Perry *v.* Sidney Phillips & Son *(1982)*

Liability in contract and tort

The chief protection against claims made for negligence in respect of professional work lies in the skill and knowledge which are acquired and maintained by continuous technical study and practical experience. The law requires a standard of care and skill which would be displayed by a reasonably competent person who has the normal skill associated with the profession in question. The law does not require all members of a profession to exhibit the very highest degrees of skill which are to be found among the most eminent practitioners, and it will not normally find a professional man or woman liable for a mere error of judgment on a difficult point.

The standard of care required is, therefore, that standard which a Court would expect to be displayed by a reasonably competent surveyor undertaking surveys and valuations of buildings. If the professional institution issues guidelines for the conduct of this type of work and if the surveyor fails to comply with those guidelines, then this will not, of itself, justify a claim for negligence. However, failure to comply with such guidelines or recommendations may make it more difficult for the surveyor to defend the position should it be necessary to do so in Court.

The standard of care required from all practitioners in a profession is, nevertheless, an exacting one. The law requires that if a surveyor puts himself or herself forward to clients as being qualified and fitted to survey buildings, then he or she should in fact be so qualified and fitted, and capable of achieving reasonable standards. If he or she is unable to carry out particular instructions in cases which involve matters beyond his or her capabilities, then those instructions should be declined.

There is little excuse for a professional man or woman who commits a gross error when advising a client about a particular property. A gross error resulting in substantial and avoidable financial loss to the client is a situation which must be avoided. Surveyors will not commit gross errors if they have studied this volume and conducted themselves in accordance with the guidelines and recommendations of their professional institutions, and have carried out careful and thorough inspections and carefully checked their reports before despatch.

A client may, of course, buy or lease a building, or lend money on it as security and suffer a financial loss in the future due to some unforeseen event, the nature of which no reasonably competent surveyor, carrying out specific instructions of a particular kind, could be expected to foresee. The surveyor may have a clear conscience that the loss suffered, while obviously regrettable, was not due to negligence. The only protection for a client against events of an unforeseen nature is insurance in so far as insurance may cover the type of loss involved. It must be emphasised that a surveyor's report is not an insurance policy, or a guarantee, and clients should not regard it as such.

It is hoped, therefore, that no reader who is a practising surveyor will ever be in the position of having committed a gross error. There may, however, be a claim from an aggrieved client in respect of some defect in a building or advice given in a report. There may be nothing of substance in the client's claim at all, or the matter may be arguable; the surveyor may find that he has acted for a client who is a 'professional' litigant who may make a habit of engaging in legal disputes of one kind or another. The purpose of this chapter is to discuss in a general way the present state of the law relating to professional negligence and associated liabilities in England and Wales by reference to the facts of specific cases.

Surveyors should ensure that their liability for professional negligence is covered by a policy placed with a reputable insurer and that the sum insured is adequate. Other insurances should be maintained, depending upon the nature of the surveyor's practice, covering public liability, and personal and employee liability. Professional indemnity (PI) insurance is compulsory for members of the Royal Institution of Chartered Surveyors (RICS). The sums insured should be subject to regular review in the light of the changes in property values and building costs generally and the nature of the surveyor's practice in particular. The temptation to under-insure in order to save premium costs should be resisted.

A legal liability which may arise will do so, in English law, under the law of contract where there is either a direct contractual relationship between surveyor and client or a contractual relationship which is held to include a third party; it may also arise under the law of torts where there is no direct contractual relationship but where the law provides that a duty to care exists. The law in both areas has continually evolved since the Second World War due to the establishment of a series of important legal precedents in certain leading cases. The potential liability of a surveyor towards clients and third parties is now much greater than would have been thought possible 40 years ago. In addition, certain limitations have been placed upon the extent to which liability may be avoided by the use of exclusion clauses, particularly by the Unfair Contract Terms Act 1977.

Important and interesting cases

In order to convey the flavour of this process of evolution a number of leading and recent cases are now summarised where the Court judgments may be of interest to surveyors, having particular regard to cases which indicate the Court's reaction to particular circumstances and interpretation of the standard of care required in specific cases. The Unfair Contract Terms Act is also discussed.

Morgan v. Perry *(1973) 229 EG 1737*

This was a case of negligence heard on 16 November 1973 before Judge Kenneth Jones in the Queen's Bench Division of the High Court, and it comprised a claim by Mr Morgan against a surveyor Mr Perry in respect of serious faults which developed in a house called 'Samarkand' at Cleeve Hill, Cheltenham.

Mr Morgan had obtained a report from Mr Perry and subsequently purchased the house. The claim was therefore a claim for professional negligence arising out of a contract between the parties, it being alleged that the defendant failed to exercise reasonable care and skill in and about the survey of the house, and the preparation of the report. These allegations were found proved.

The house was built upon a steeply sloping hillside which, in fact, was so steep that the house was of split-level construction with the living rooms at first floor level at the rear and ground floor level at the front, and the bedrooms at ground floor level at the rear and forming a basement at the front. The house had been built in 1964 and the defendant surveyed the house in July 1968, issuing his report the following day. Under 'Condition external' he reported: 'The structure is sound and there is no evidence of settlement'. Under 'Condition internal' he reported: 'There are some hairline plaster cracks. These are of a minor nature.' Under the heading 'Opinion' he reported: 'We are of the opinion that this property is structurally sound. There should be no undue maintenance costs in the foreseeable future.'

The Judge commented that the report was a remarkable document; apart from the minor hairline cracking to plaster in unspecified locations, no defect of any kind was referred to. The defendant had given the house a completely clean bill of health and the plaintiff had proceeded with the purchase. However, while laying the drawing room carpet the plaintiff had noted that the floor was not level. Moreover, after taking occupation he had noted certain cracks in the external walls. Lacking any expert knowledge the plaintiff, at that time, had not appreciated the significance of the sloping floor or the cracks.

In May 1969 the plaintiff applied for a mortgage, and in June 1969 a survey had been undertaken on behalf of the Halifax Building Society. The Halifax surveyor found evidence of settlement, and eventually, having failed to obtain subsidence insurance, the Halifax declined to advance any money on the security of the house. In due course bore holes were sunk and advice taken. A firm of consulting engineers reported in January 1972.

Engineers reported that the topsoil to a depth of 3.3 m was slipping on the consolidated clay beneath. The slip was taking place in the three masses: one mass was slipping rotationally; one mass was slipping so that the garage was tilting towards the house; and a general movement down the hillside was taking place. The engineers advised demolition of the house, slope stabilisation work, special foundations and a new house, this being the only way to produce a house on the site with firm and lasting foundations.

The Court decided that it was fundamental to establish the visible condition of the house in July 1968 when the original survey took place. Twelve witnesses were called to give evidence, some by the plaintiff and some by the defendant. The original owner, and vendor to the plaintiff, was called to give evidence and the Judge found that he was less than frank with the Court regarding defects which apparently arose during his period of ownership, and some filling of cracks which had been undertaken prior to sale. Whether or not the vendor had deliberately covered up the defects, the Judge found that the effect of crack filling, replastering and wallpapering which had been undertaken was the same as that of a deliberate cover-up job with the intention of making the house saleable. The Judge decided that the defendant, Perry, did not in fact observe any defects when he carried out his survey, but he found it impossible to rely upon his evidence that no defects were observable.

The Judge found that the defendant had failed to notice repointing, or the shape and extent of various cracks, repairs to plasterwork, doors not closing properly, over-painting and so forth. He commented that, human nature being what it was, there would be instances of vendors taking steps to cover up defects, and that was at least part of the reason why a prospective purchaser employed a surveyor. Considered against the fact that the house was built on a steeply sloping site with known potential dangers, the surveyor should have undertaken a close examination of the house, which would have revealed the slopes to the floors. Evidence was given that the highway outside the house had been subsiding for many years, and the surveyor should have looked at the highway, which would have aroused his suspicions and led him to make enquiries with the highway authority. The Judge found that the defendant committed a breach of his contract with the plaintiff and provided him with a report which was so misleading as to be valueless. Accordingly, an award of substantial damages plus costs was made.

Leigh *v.* Unsworth *(1972) 230 EG 501*

A case rather similar to *Morgan* v. *Perry*, but with a different outcome, was heard before Judge Everett QC in the Queen's Bench Division of the High Court on 21 November 1972. This comprised a claim by Mr Leigh against a surveyor Mr Unsworth in respect of faults in a house called 'Greenacres' at Swallowfield, Berkshire.

The defendant surveyed the house in April 1965 and the plaintiff subsequently purchased. The writ against the defendant was not issued until 5 years and 51 weeks after the date of the survey, a matter which the Judge described as regrettable, especially as a further delay then ensued before the matter was finally brought to trial, so that 7½ years elapsed between the date of the survey and the date of the trial.

The house had apparently been built in about 1960 to a very poor standard, especially in respect of the main defect which was that the outer walls rested on conventional foundations but the internal walls rested on a concrete raft supported by a backfilling of loose clay and other unsuitable material, rather than conventional hardcore. It was agreed at the trial that the manner in which support was provided for the internal walls was lamentable building practice. The issue was whether, at the time of the survey, there were present sufficient indications of the existence of settlement of the internal walls and raft, such as to make a competent surveyor advise his client of the problem.

The Judge held that the duty of the surveyor in these circumstances falls into two areas. The first is actual observation, and the inferences to be drawn from the results of that observation. The second is just what is required of him by way of information, put quite broadly, to his client. In the report the defendant had stated: 'Generally the property is soundly constructed and no major defects were found. The works of maintenance and repair are generally maintenance works due to original shrinkage and drying out.' He also stated that the chimney stacks had lead flashings, whereas in fact they had cement fillets which subsequently leaked. The report stated that: 'The roof was found to be strongly framed with good-quality timbers of ample-sized scantlings' and 'the ceilings were covered with fibreglass except for a proportion of the roof which was not so treated.' The defendant claimed that the main roof was poorly constructed, in that there was a complete lack of vertical hangers, and purlins were not properly strutted having individual lengths butt-jointed and cantilevered. Further that, apart from one small area, the ceilings were not in fact insulated. The defendant had rendered an account for builder's charges in connection with drains tests, but, it was alleged, had failed to discover a massive leak in a drain under the building.

Having moved into the building the plaintiff discovered various defects, including the fact that the kitchen floor had sunk between 12 and 18 mm. During the course of the trial evidence was given from each side as to the condition of the premises as it had developed, which was agreed to amount to a serious settlement, and the condition of the premises as would have existed at the time of the survey. The outcome was that the Judge found that three minor errors in the report had been made: the poor construction of the roof timbers; the lack of lead flashing to the stack; and the poor level of insulation.

As to the substance of the case, which was the settlement of the internal raft and walls, the Judge found that the plaintiff's case had not been made out, and an award for the three minor matters only, plus some costs, was made. The leakage of the drains was considered to have arisen after the date of the survey, and the defendant's evidence, that the drains passed a test at the time of the survey, was accepted.

Clearly, if the three minor errors in the report had not been made the outcome of the case would have been to vindicate the surveyor, in the circumstances described. The significance of the fact that the writ was served 5 years and 51 weeks after the date of the survey lies in the fact that an action for breach of contract must be commenced within 6 years of the date of the alleged breach. In order to be able to defend an action for breach of contract all records relating to the matter must be retained for at least 6 years. In respect of actions in tort the time limit may, in effect, be longer since such actions may be commenced within 6 years of damage resulting to the building (see *Pirelli General Cable Works Ltd* v. *Oscar Faber & Partners* [1983] 1 All ER 65, [1983] 2 WLR 6). For this reason it is suggested that all surveyors' records dealing with surveys and reports be retained, in their entirety, without time limit.

Hedley Byrne & Co. Ltd *v.* Heller & Partners Ltd *[1963] 2 All ER 575, [1964] AC 465*

In Hedley Byrne the House of Lords introduced a new principle into the common law which was that a liability arose in tort between the maker of a statement of fact and opinion, and a person who may rely upon that statement of fact or opinion, even if there is no direct contractual relationship between them. Prior to Hedley Byrne the Courts had restricted liability in tort to actions, rather than the spoken or written word, and restricted the right to compensation for damage to that occasioned by physical damage rather than financial loss.

Hedley Byrne established a new type of liability in tort as being that for the spoken or written word giving rise to financial loss where the writer or speaker and the person who relies upon the advice have a relationship of proximity or neighbourhood which gives rise to a duty to care.

Dutton *v.* Bognor *Regis Urban District Council [1972] 1 All ER 462, [1972] 1 QB 373*

In *Dutton* v. *Bognor* the question of liability in tort was further considered, and extended. In this case the plaintiff, Mrs Dutton, purchased a house in 1960 not knowing that it had been built on made ground, in fact an old rubbish tip. It became clear, shortly after the purchase, that the house was unstable, and the plaintiff sued the local authority for negligence and breach of statutory duty in respect of the inspection of the house when under construction pursuant to their powers under the Public Health Acts and local bye-laws (since superseded by the Building Regulations). The trial Judge found that the local authority surveyor had been negligent in not noticing evidence of rubbish in the foundation trenches before they were covered-up.

Murphy *v.* Brentwood UDC and DoE *v.* Thomas Bates and Son *[House of Lords – 26 July 1990]*

In these two important later cases the position established in *Dutton* v. *Bognor* [supra] and confirmed in *Anns* v. *Merton LBC* [1978] was overruled. These cases were found to have been wrongly decided as regards their statutory functions. The result of these most recent cases appears to be that local authority liability extends to physical injury to persons or property on account of defective works but not the defective buildings themselves.

Yianni *v.* Edwin Evans & Sons *(a firm) [1981] 3 All ER 592, [1981] 3 WLR 843*

In the High Court Mr Justice Park ruled that there was a sufficient relationship of proximity between the surveyor to a building society and the borrowers that the borrowers could sue the surveyors in tort, notwithstanding disclaimers of liability used by the building society.

In 1975 Mr and Mrs Yianni, as sitting tenants, were offered a house in North London at a price of £15 000. They applied to the Halifax Building Society for a mortgage of £12 000 which was an advance of 80% needed in order that they could buy. The Halifax instructed their surveyors, Edwin Evans & Sons, to prepare a pro forma report and valuation of the house in the normal way. In their offer of a mortgage advance the Halifax enclosed an explanatory booklet which included a statement that the Building Society accepted no liability for the accuracy of the valuation, and that the valuer's report was confidential information used by the Society to determine whether or not an advance should be made. It was stated that if the buyers required a survey they should instruct an independent surveyor, and indeed this was recommended. The Yiannis did not read the booklet, and the report by Edwin Evans & Sons did not reveal any faults. In due course the Halifax lent £12 000 as an 80% advance indicating that their valuation was £15 000.

It transpired that some cracks had apparently appeared in the house before it had been offered to the plaintiffs, and the then owners had discovered subsidence. Some repairs and redecorations had been carried out prior to the survey. After the purchase further cracks appeared and were diagnosed as resulting from subsidence; it was found that the end wall of the house needed rebuilding, and other walls needed underpinning. By 1978 the estimated cost of repairs had risen to £18 000. Edwin Evans admitted negligence in failing to notice that the building had been subject to past subsidence, in failing to take proper steps to have reasons for subsidence investigated, and in reporting to the Halifax that it was suitable security for an 80% loan. They claimed, however, that they did not owe any duty to care to the plaintiffs, with whom they had no contractual relationship and with whom there was no liability in tort.

For their part the plaintiffs contended that Edwin Evans owed a duty of care to them since they relied upon the Halifax offer as indicating that the property was worth £15 000. The Judge agreed that a relationship of proximity existed such that, in the reasonable contemplation of the defendants, carelessness on their part might be likely to cause damage to the plaintiffs.

During the course of the hearing evidence was given that perhaps only 10% of house purchasers obtain an independent surveyor's report of their own, the remainder relying on the building society surveyor or their own inspections as being sufficient indication that the property is sound and worth the price. In these circumstances the Judge held that the plaintiffs were not liable for contributory negligence in failing to obtain their own report, and the defendants were found liable for substantial damages. On the facts of this case at least the Courts have decided that a purchaser is right to place some reliance upon the building society report as an indication of the condition, and value, of the property. This decision in 1981 sent a shockwave through the previously rather sleepy world of the Building Society valuer and led to a spate of similar claims.

Donoghue (or McAlister) *v.* Stevenson *[1932] All ER Rep 1, [1932] AC 562*

It will be seen from the summaries of the foregoing Hedley Byrne, Dutton and Yianni cases that the range of persons to whom a surveyor may be held responsible in tort is wide, and has progressively widened in recent years. The process by which the Courts can make new law in this way is termed the Doctrine of *Stare Decisis*, by which points of law are determined by the Courts so as to establish precedents, the decisions of the Appeal Courts having greater weight than those of the High Court, with the decisions of the House of Lords, the premier appellate court (the European Courts excepted), having the highest. A new Supreme Court will shortly replace the House of Lords.

In *Donoghue* v. *Stevenson* Lord Atkin described the range of people to whom a duty to care is owed, under the law of torts, in the following terms: 'The rule that you are to love your neighbour becomes, in law, you must not injure your neighbour and the lawyers' question "Who is my neighbour?" receives a restricted reply. You must take reasonable care to avoid acts or omissions which you can reasonably foresee would be likely to injure your neighbour. Who then, in law, is my neighbour? The answer seems to be – persons who are so closely and directly affected by my act that I ought reasonably to have them in contemplation as being affected when I am directing my mind to the acts or omissions which are called in question.'

Bolam *v.* Friern *Hospital Management Committee [1957] 2 All ER 118, [1957] 1 WLR 582*

In this case the standard of care required from a professional man was described by McNair J. as 'The test is that of the ordinary skilled man exercising and professing to have that special skill. A man need not possess the highest expert skill: it is well-established law that it is sufficient if he exercises skill of an ordinary competent man exercising that particular art.'

Rona *v.* Pearce *(1953) 162 EG 380*

In this case the extent to which a surveyor should describe and comment upon aspects of a building and the action he should take in respect of those matters which could not be inspected or confirmed was considered by Hilbery J., who said: 'It is highly important to ordinary members of the lay public that a surveyor should use proper care to warn them regarding matters about which they should be warned on the construction or otherwise of a piece of property and that they should be told what are the facts. If surveyor cannot ascertain any fact he should say he was unable to ascertain it and therefore unable to pass any opinion on it.'

Perry *v.* Sidney Phillips & Son *(a firm) [1982] 1 All ER 1005, [1982] 3 All ER 705, [1982] 1 WLR 1297*

This case, which was a claim arising out of a negligent survey, was subject to hearings in the High Court and subsequently in the Court of Appeal. The case is of interest because of the facts, and because the issue of the quantum of damages was considered; the issue on damages being whether these should be assessed on the basis of diminution of value, or cost of necessary repairs, and the date at which damages should be assessed.

In 1976 Mr Perry bought a cottage in Worcestershire for £27 000, having first obtained a report from surveyors Sidney Phillips & Son in which the latter reported that the cottage was generally in order and a reasonable buy at the price. Subsequently Mr Perry discovered a number of defects and commenced an action for breach of contract and negligence.

In the High Court the main allegations were either conceded by the defendants, or upheld by the Judge. On the understanding that Mr Perry intended to remain in the house for the foreseeable future the High Court Judge awarded damages for (i) the cost of repairing defects which the surveyors had negligently overlooked; (ii) compensation for the inconvenience and discomfort anticipated while the remedial works were being carried out and (iii) compensation for distress, discomfort and other consequences of living in a house in defective condition. The High Court Judge further held that repair costs were to be assessed at the date of judgment, rather than the date of the breach of contract since the defendant was unable to undertake the works at an earlier date due to lack of funds. The High Court judgment on negligence was accepted by the defendants, but they appealed to the Court of Appeal on the issue of the assessment of damages.

The facts, which were not at issue in the Appeal Court, were set out in the High Court judgment ([1982] 1 All ER 1005). The defects in the property, in respect of which a lack of care in the undertaking of the survey and preparation of the report was established, included a number of defects in the roof, a blocked doorway, lack of weatherproofing to windows, a chimney stack lacking proper support, a bowed and bulging wall, water penetration to the interior and various minor defects including a loose basin and loose wall tiles. In addition, the cottage had septic tank drainage which was unsatisfactory and not functioning correctly; the drainage system was a nuisance and generated unpleasant smells.

Unfortunately, by the time this case reached the Court of Appeal the plaintiff's circumstances had changed and he had sold the cottage; it therefore followed that it was no longer possible for the plaintiff to argue the merits of 'cost of repair' as an alternative to 'diminution in value'. The original award of damages based upon cost of repair was made only on the understanding that the plaintiff intended to reside in the house as his home for the foreseeable future.

The Appeal Court reduced the damages to a figure equal to the difference between the price paid for the house, and its true value, as at the date of the purchase, plus interest as some compensation for inflation and 'vexation'; compensation for vexation being a modest recognition of the anxiety, worry and distress arising from the physical consequences of the breach. Damages were to be assessed by the official referee. The surveyors' appeal was therefore allowed on the principle that damages are assessed on the basis of diminution in value, and as at the date of the breach. Two members of the Appeal Court were careful to emphasise that the changed circumstances prompted the decision, and that 'cost of repair' was, at the time of the Appeal Court hearing, no longer arguable. The third member, Lord Denning MR, was however prepared to venture a more general opinion of the position, and he said: 'Damages are to be assessed at the time of the breach, according to the difference in price which the buyer would have given if the report had been carefully made from that which he in fact gave owing to the negligence of the surveyor . . . The buyer is not entitled to remedy the defect and charge the cost to the surveyor. He is only entitled to damages for breach of contract or for negligence . . . So you have to take the difference in valuation . . . In other words, how much more did he pay for the house by reason of the negligent report than he would have paid had it been a good report?'

Lord Denning added that, while damages would be assessed at the date of the breach, they would carry interest, so that by this means some protection against inflation was provided.

The Appeal Court was not required to assess damages. However, the principle of damages for vexation was confirmed, and some indication given that such sums as may be awarded under this heading should not be large; Lord Denning referred to 'modest compensation', Oliver LJ suggested that such damages would be 'not very substantial', and Kerr LJ confirmed that vexation damages were limited and not intended to deal with 'tension or frustration of a person involved in a legal dispute . . . such aggravation is experienced by almost all litigants'. Vexation damages were intended to deal only with the physical consequences of the surveyors' breach of contract.

Eames London Estates Ltd *v.* North Hertfordshire District Council *(1981) 259 EG 491*

In the event that defects arise in the construction of a building, a number of parties may be at fault, apart from the surveyor who may have inspected the building for the purchaser. Rules governing the extent of liability have been stated in a number of post-war legal actions, in addition to those already mentioned. On occasion, local authority building control, architects, consulting engineers, builders, developers and surveyors have all been found liable of breach of duty to care, breach of statutory duty or negligence. For a broad view the reader is recommended to study the judgements in *Anns* v. *Merton London Borough Council* [1977] 2 All ER 492, [1978] AC 728; *Batty* v. *Metropolitan Property Realisations* [1978] 2 All ER 445, [1978] QB 554; *Acrescrest Ltd* v. *W. S. Hattrell & Partners* (a firm) [1983] 1 All ER 17, [1982] 3 WLR 1076 and *Eames London Estates Ltd* v. *North Hertforshire District Council* (1981) 259 EG 491.

In Eames the case concerned an industrial building which was designed by an architect, built by a firm of contractors for a firm of developers, and subsequently let with the freehold reversionary interest sold to investors. The building was erected partly on made ground subject to past landfilling, and some cracks appeared in 1971, but it was not until 1976 that the cracking caused such alarm that expert advice was sought by the tenants.

Eventually a total of three plaintiffs took action against four defendants, the defendants being the architect, the local authority who passed the foundations, the developers and the builders. Judge Fay held that all four defendants were liable to pay damages. The architect was considered to bear a high degree of responsibility, since the building, and its foundations, were his design; the fact that the local authority had passed the design was no excuse. The architect was found liable to pay 32½% of the total damages awarded. The developers were held liable for 22½% of the damages having regard to their apparent knowledge of the site conditions, and lack of suitable instruction as to foundation design. The builders were held 22½% liable, despite evidence that they had found fill material in the trenches and reported this to the architect, and proceeded only on the architect's instructions. The local authority were also held 22½% liable for failing to take sufficient steps to confirm that the foundation design was suitable.

Eames is of interest because of the manner in which liability was apportioned among the negligent parties. The developers could not evade liability by pointing to the fact that they had consulted an architect; they were still liable to ensure that the land was suitable for the foundations proposed (following *Batty* v. *Metropolitan Property Realisations* (supra)). The builders were unable to plead successfully that they were acting on the instructions of the architect and building inspector from the local authority in continuing to lay foundations in bad ground.

Eames also touched upon the question of the time from which the period of 6 years prescribed in the limitation Act for actions in tort or contract would run in circumstances where progressive crack damage was appearing in the building. It was held that this period ran, for the purpose of an action in tort, from that time when reasonable skill and diligence would have enabled the plaintiffs to have detected not merely the symptoms (cracks) but also the nature of the underlying cause (differential settlement). In Eames the plaintiffs' case against the defendants was held to be within the necessary time limit, having regard to the evidence given as to the nature of the structural movements involved and the dates at which crack damage became apparent. Interestingly, the architect claimed to have known in 1972 that differential settlement was probably the cause, this being some time before the plaintiffs were held to have appreciated this fact; it was held, however, that the architect could not rely upon his own knowledge to defeat a claim made against him on the grounds that the claim was therefore out-of-time.

Balcomb *v.* Wards Construction (Medway) Ltd; Pethybridge *v.* Wards Construction (Medway) Ltd. *(1981) 259 EG 765*

In the Balcombe and Pethybridge cases two claims were heard together, both being claims by the purchasers of houses against the builders and their consulting engineers arising out of cracking in newly-built houses resulting from heave damage, the buildings having been erected with shallow foundations on a London clay site where a number of trees had been removed prior to commencement of construction.

Evidence was given that, following the removal of trees, heave damage can last for many years, perhaps as many as 10 years, and that the shallow foundations used at a reputed depth of 1.05 metres were unsuitable. Engineers had advised the use of foundations at this depth, suggesting that it would be deep enough to avoid damage from any change in the condition of the clay. The houses had been built in 1972, and in 1974 substantial repairs were undertaken; this did not cure the problem and it was agreed, for the purpose of the legal action, that the cost of proper repair exceeded the value of the houses.

The builders, Wards, admitted liability in contract to the purchasers and agreed the quantum of damages during the course of the trial. The consulting engineers also agreed the quantum of damages if they were liable to the purchasers or the builders, but they disputed the issue of liability. The Judge, Sir Douglas Frank, held that the engineers were liable to the builders for breach of contract in failing to exercise professional skill, and also liable to the builders in tort for breach of duty to care. The Judge also held the engineers liable directly to the purchasers in tort following *Bolam* v. *Friern Hospital Management Committee* (supra) and *Batty* v. *Metropolitan Property Realisations* (supra).

The Judge considered, on the evidence, that in 1971 when the engineers took soil samples and gave their advice to the builders a competent engineer, encountering London clay, would have made enquiries to discover whether or not there had been trees on the site and, finding that there had been, would have caused moisture content and plasticity tests to be made on the clay. It was considered that the results of such tests would have necessitated different advice from that actually given in that deeper, and special foundations would have been required to accommodate future clay swell.

Interestingly it was held that the builders were not themselves liable for negligence; the Judge considered that the builders discharged their duty by employing engineers to advise them, in this respect differing from the Eames case (supra) where the employment of an architect was not considered sufficient to absolve either the developer or the builder from a measure of responsibility for defective construction. Having regard to the fact that Wards were not themselves qualified to carry out the site investigations the Judge found it impossible to think of any other way that they could have discharged their duty other than to rely upon the consulting engineers' advice; accordingly no contributory negligence by the builders was found, and the full burden of liability for the defective foundations rested with the engineers.

Surveyors are advised to foster a relationship with a firm of consulting engineers, and to refer to them for advice on structural matters in many situations (see Chapter 3). It is suggested that, in view of the possibility that a shared liability may arise for advice given, all requests for assistance from structural engineers should be made in writing stating the specific services required and requesting that engineers themselves should advise on the nature and extent of any testing or sampling they require to provide complete advice and that any limitations involved be carefully explained, in writing, to the client.

McGuirk *v.* Homes (Basildon) Ltd *and* French Kier Holdings Ltd (Estates Times 29 *January 1982*)

In this case a claim was made by the purchasers of a house built with mortar found to consist of only one part Portland cement to thirteen parts of sand. The very weak mortar was stated by a structural engineer, called by the defendants, to have complied with building bye-laws applicable at the time, and with BSI CP 111:1970, and the defendants denied any breach of duty.

The Judge found differently, in that the duty of the defendants was 'to build in a substantial and work-manlike manner, and not merely to comply with the bye-laws'. The house had been built in 1960 and the local bye-laws applicable at the time were those of Basildon Urban District Council dated 1958. After 1965 the bye-laws were replaced by the Building Regulations in England Wales (outside inner London), and BS Code of Practice CP 111 has been revised, so that it is thought unlikely that 1:13 mortar could be defended today on the grounds that it complied.

The defect was not discovered for some years. In 1978 it was noticed that smoke was escaping from the chimney stack through the mortar joints, and a local builder was asked to investigate. He found the mortar between the bricks being blown away by the wind and many bricks loose so that they could be lifted out. Further investigation by the plaintiff revealed that the problem was quite general and applied to all the brickwork in the house.

The house has been purchased in 1960 for £3165. The eventual award of damages for diminution in value and loss of amenity was £6684.92. Concluding his judgment Judge Leonard said that the case had been a misfortune for both sides, in that the defendants found themselves saddled with an unexpected liability over 20 years after they had sold the house, and he added 'I regard this as a misfortune for them because I think it is most improbable that anyone in authority intended an unduly weak mortar mix to be used, or suspected that it was being used. The saving in expense would be minimal. It is no part of my function to speculate on the explanation. It is enough for one to say that nothing in the case has come to light suggesting any deliberate intention on the part of the defendants to build otherwise than in accordance with the contract. No word has been said suggesting that the rest of the house was not properly and soundly built'.

It is worth emphasising that a surveyor inspecting a building under construction can undertake only spot checks on materials and workmanship. The modern practice of dispensing with an on-site clerk of works (who represents the client or purchaser rather than the contractor), and the impracticability of undertaking continuous on-site supervision, means that, no matter how frequent and diligent are the surveyor's inspections, they cannot prevent deliberate breaches of good building practice by the workforce occurring between site visits. If workmen on-site wish there are many opportunities for them to undertake work badly, cover up the evidence and avoid detection. Some indication of the overall quality of the workmanship and the skill and honesty of the tradesmen on-site can, however, be gleaned from meeting and talking with the workforce (see Chapter 14).

London & South of England Building Society *v.* Stone *(1981) 261 EG 463*

Mr Stone prepared a pro forma building society report on a house in Wiltshire and valued it at £14 850, the society then lent £11 880 to the purchasers Mr and Mrs Robinson. The house subsided, and remedial works eventually cost £29 000. The society decided to waive their rights under the mortgage deed against Mr and Mrs Robinson and proceeded against Mr Stone for £29 000 plus costs. Mr Justice Russell did not award the cost of remedial works as damages, but adopted a lower figure of £11 880 as being the difference between the stated value, and actual value of the property at the time of the report, and further deducted £3000 from this because he considered that the society had failed to mitigate their loss by waiving their rights against the borrowers. A net award of £8880 plus some costs was made.

The judgment recorded that the surveyor, Mr Stone, was a frank and honest witness who displayed a professional approach to the allegations made and emerged from the witness box with credit. He described his inspection, lasting about an hour, during which he had not seen anything untoward.

The case against Mr Stone was that he should have noticed an important crack, even allowing for some cosmetic remedial works which had been undertaken, and that he should have detected rotational movement in an outhouse, and that he should have noted sticking doors and cracking inside the house. Prior to his survey the then owner had apparently made an insurance claim for subsidence damage, which had been resisted by the insurers on the ground that it originated before they came on risk. The house had been built in about 1961 on sloping ground and on the site of an old quarry. Records existed in the form of a surveyor's report and photographs prepared for the insurers, indicating the condition in February 1976. The inspection by Mr Stone was in May 1976, so that considerable evidence existed of a major structural fault as at the date of Mr Stone's inspection.

The reader will not need reminding by now that I have emphasised in this book the desirability of acquiring local knowledge about old quarry workings and the like, and the higher risks of structural damage likely to arise with buildings erected on sloping, as opposed to level, sites which should place a surveyor on guard.

The Judge noted that attempts had been made by the vendor to better his property and that this would have resulted in some covering up of crack damage. It was felt that evidence of past internal damage was not available to be seen when Mr Stone inspected. The Judge was, however, driven to the conclusion that the crack in the exterior of the structure, and the rotational movement in the outhouse, were available to be seen and that, for whatever reason, Mr Stone had failed to detect them. The Judge found this to be an isolated lapse on the part of a surveyor of considerable expertise and experience, but that liability was plainly established.

The Judge commented upon the nature and extent of a building society survey; that it was not a full structural survey but merely a survey for building society valuation purposes. Nevertheless, the surveyor in this instance had failed to exercise the degree of care and skill expected of a reasonably competent surveyor carrying out the type of examination that he was required to carry out on behalf of the building society.

Fryer v. Bunney *(1982) 263 EG 158*

An action for negligence and breach of duty to care was brought against a surveyor arising out of serious dampness found in a house due to leakage within solid floors from defective small-bore central heating pipework. The defendant surveyor, Mr Bunney, reported that the house had been checked with a moisture meter and that no reading of dampness had been found. When the purchasers took possession they soon discovered serious dampness, which was confirmed by measuring the loss of water from the central heating header tank. It became necessary to excavate the hall floor, remove defective pipework and renew joints. The judgment went in favour of the plaintiffs and the defendant surveyor was held to have been negligent in not noticing or recording the dampness. Damages of £5346 were assessed, comprising cost of repairs and decorations, reduced value of shrunken carpets, and an amount for vexation arising out of the plaintiff's inconvenience and distress (the cost of remedying the defects being presumably equivalent to the diminution in value of the property).

Problems associated with small- and micro-bore pipework embedded in solid floors are well recognised. It is essential that such pipework be tested for leaks under pressure before screeds are applied; frequently this is not done. In *Fryer* v. *Bunney* the evidence was that soldered joints had not been properly made, and pipes were merely a push fit (indicating, among other things, the contractor's negligence). It is also essential that lagging be properly provided to permit thermal expansion and contraction. Even well-made joints can fail if the pipework is screeded in solidly and unable to respond to thermal movements. From a surveyor's point of view such pipework is inaccessible on subsequent survey, and the only practicable tests are of moisture levels in surrounding walls and floors. Unfortunately for Mr Bunney it was held that he had not undertaken such testing to the standard and extent required in the circumstances.

Fisher v. Knowles *(1982) 262 EG 1083*

The purchasers claimed against their surveyor some 4 years after the purchase, particularising 14 items as being alleged defects; the judge found that the surveyor had failed to warn his clients of only two matters, being some wet-rot to window frames and defects to ceiling joists.

Instructions arose because the surveyor was to inspect the house for the building society, and the purchasers approached him with a request that he also provide them with a private report and valuation. The Judge found that these instructions amounted to something more than a bare valuation report, but something less than a full building survey. The RICS Practice Note (which was not published when the allegedly negligent inspection was made) now recommends that if something less than a full building survey is to be carried out it is important that the true nature and extent of the obligation being assumed by the surveyor is understood by, and agreed with, the client.

The Judge emphasised that the burden is on the plaintiffs to prove that, on the balance of the probabilities, the defects were of a kind to which the defendant should have drawn attention. Making the best he could of the evidence the Judge awarded £500 damages for diminution in value of the house, resulting from the two alleged defects found to have existed at the time of the inspection and missed by the surveyor. A claim for a further 12 items was unsuccessful.

Wet-rot to softwood window frames is a known hazard, and the defendant surveyor was in the habit of testing joinery with a penknife; he considered that he would have done so in this case. The Judge found, on balance, that there had been wet-rot to the window frames, and that for some reason the surveyor had failed to detect it. He also felt that the surveyor had failed to notice deflecting ceilings under a bathroom floor due to the poor fixing of a water closet pan, which would have placed him on guard.

The author would recommend that particular care be taken in inspecting exterior painted softwood for signs of rot, and also whenever possible in checking conditions under water closet pans resting on timber floors, since rot to timbers from leaking seals will commonly be found.

Treml v. Ernest W. Gibson & Partners *(1984) 272 Estates Gazette 68*

Ernest Gibson, a member of the Incorporated Society of Valuers and Auctioneers, was held liable for substantial damages in this case which illustrates how damages can mount up if serious building defects are found. The facts, of considerable interest to practitioners, were briefly as follows.

Mrs Treml had agreed to buy No. 9, Keens Road, Croydon, Surrey in June 1979 for £21 000, subject to survey. Mr Gibson was instructed and reported on this small-terraced house finding no obvious defects but advising some treatment of woodworm and rewiring; Mrs Treml bought the house and moved in.

Unfortunately Mr Gibson had failed to notice that the house lacked firewalls in the roof void and had a modern concrete tile covering which was much heavier than the original slate covering. The roof trusses were thrusting against the front and rear walls. In time bulges appeared in those walls as the thrust increased until shoring and structural support was needed to prevent collapse – in October 1982 the local authority actually issued a Dangerous Structure notice.

Mrs Treml asked Mr Gibson to return to the house in 1980, 1981 and 1982 because of various problems but Mr Gibson failed to notice the fundamental defect. Eventually Mrs Treml took expert advice and the nature of the problem became apparent. By the time the case came to court the defendant conceded negligence and the only question to be determined became that of damages.

Mr Justice Popplewell said that following the decision in *Perry* v. *Sidney Phillips & Son* (supra) the proper method to assess damages was to calculate the difference in value between what the plaintiff actually paid and what the real value of the house was, and to add interest plus any special damages for vexation and inconvenience, and the temporary costs involved which arose in this case because the plaintiff would have to move out of the house for a time whilst it was repaired. Evidence that the real value was £8000 was accepted, leaving a sum of £13 000 as damages for the negligent survey plus interest and vexation damages.

Mrs Treml had also claimed for the cost of employing engineers, surveyors and builders to arrange emergency support and advise on the nature of the defect. The builders charged £756, the engineers £1255 and the surveyors £115. The judge did not allow engineers costs apart from £50 relating to the report dealing with the question of the defendant's negligence, the reason being that having assessed damages in relation to the market value of the property it would be unfair also to add further damages which were in the nature of repair costs as this would result in the plaintiff securing compensation in excess of the actual damage suffered.

As an additional complication it transpired that grant assistance was available under the 1974 Housing Act from the local authority in respect of a proportion of repair costs. The Judge held that the possibility of grant aid being available was irrelevant to a claim where the measure of damage was the difference in value, a point strongly argued by counsel for both sides.

Further damages were awarded for a number of incidental costs including £3850 for the cost of 70 days at a hotel at £55 per day for Mrs Treml and her family whilst the work was done, plus £1000 for meals and £50 for laundry. A further £1248 was awarded for the costs of furniture removal and storage, and accommodation for Mrs Treml's cat. £1250 for vexation and inconvenience had been agreed by the parties. Total damages came to £32 750 plus interest of £8500. The judge ordered a stay on £6000 of these damages to cover the possibility that the defendant might appeal, on a point of law, that the local authority grant should have been deducted.

Damages at the end of the day were much higher than they might have been had remedial work been carried out at an early stage, but the court recognised that the plaintiff could not in fact carry out any work until she had successfully sued the defendant, because she did not have the money. Moreover, raising finance would have been difficult on her income and with no assets save this defective house. This being a terraced house there was also some disagreement initially between neighbours on the nature of the works required.

It is suggested that a surveyor should always inspect the roof void if this is accessible. A lack of party walls carried up to provide firewalls in terraced roof voids will commonly be encountered in terraces dating from the last century and this must always be commented upon. If light coverings, such as slates, are replaced with heavier coverings, such as concrete tiles, it is essential that a structural check be first made to ensure that the timbers will take the higher loads – this is often not done resulting in roof sag or roof spread. Consent under Building Regulations is generally required for work involving a heavier roof covering – but roofing contractors do not apply in many cases.

Stevenson v. Nationwide Building Society *(1984) 25 July 1984*

A staff valuer employed by the building society inspected a property which consisted of two shops, a maisonette and a flat built over a river and supported on steel girders and an infilling of concrete. The plaintiff was an estate agent who was purchasing the premises with the assistance of a loan from the society.

When the plaintiff made the application for the loan he completed an application form which included the following clauses:

> 'The inspection carried out by the Society's Valuer is not a structural survey and there may be defects which such a survey would reveal. Should you wish to arrange for a structural survey this

can be undertaken by the Society's Valuer, at your own expense, at the same time as the Society's Report and Valuation is made.

I understand that the Report and Valuation on the property made by the Society's Valuer is confidential and intended solely for the consideration of the Society in determining what advance (if any) may be made on the security, and that no responsibility is implied or accepted by the Society or its Valuer for either the value or the condition of the property by reason of such Inspection and Report.'

The Staff Valuer's report recommended a retention of £5000 out of a total advance of £39 000 until certain specified work was done to the roof, electrical system and external decorations. No reference was made to the fact that the building spanned the river and was of very peculiar construction.

The plaintiff bought the property in May 1982 and let the second shop to a butcher. In June the butcher was using the lavatory at the rear of his shop when part of the floor collapsed. A subsequent survey indicated that the building was in a dangerous condition and in particular that the entire floor construction at the rear was beyond reclamation.

The plaintiff then claimed damages against the society on the ground that their valuer had been negligent in the preparation of his report. The Judge (Mr J. Wilmers QC) was therefore required to consider just what, in practical terms, is expected of a valuer carrying out a building society inspection. He then had to go on to consider whether the disclaimer, which the plaintiff had signed, affected the situation.

On the first point the Judge held that the valuer had fallen below the standard of care and skill reasonably expected in the circumstances. His opinion was:

'I bear in mind that [the valuer] was called on to do a building society valuation and not a structural survey. Nevertheless, I have no doubt that I must hold him not to have exercised reasonable skill or care. Given the nature of these premises, I do not think that it is possible to discharge properly the duties of a valuer unless he either looks under the building himself or, if he for some reason could not do so, ensured that he had the report of some other competent person who had done so. In either event, the actual state of these premises would have been revealed. I do not need to determine the depth of water at the time of inspection. Any valuer reporting on these premises, constructed as they are, must, in my view, either brave such water as he finds or find someone else to do it.'

The Judge determined that the valuer and the building society, as his employers, were prima facie liable in respect of this negligent report. But he then went on to consider what effect the disclaimer had on the situation.

Significantly the plaintiff was an estate agent with offices in Wiltshire and Somerset. He had no experience in valuation or surveying, but he was aware of the nature of the disclaimer when he signed it and he had been given the opportunity to commission a more expensive report on the building – and declined to take up that offer. In the circumstances the judge held that the society, and their valuer, were protected by the disclaimer.

The judge considered the Unfair Contract Terms Act 1977 and said:

'It seems to me perfectly reasonable to allow the Building Society, in effect, to say to [the plaintiff] that if he chooses the cheaper alternative he must accept that the Society will not be responsible for the contents to him.'

This decision effectively confirmed that the society's valuer had been negligent in the preparation of his valuation, but by upholding the society's standard form of disclaimer it deprived the plaintiff of any remedy.

Watts and Another *v.* Morrow *(1991) 14 EG 111*

His Honour Judge Peter Bowsher, sitting as Official Referee, found Ralph Morrow, a building surveyor, negligent in carrying out a Building Survey of Nutford Farmhouse in Dorset for Mr and Mrs Watts and an award of damages was made based upon the cost of repairs (£34 000) plus an amount for distress and inconvenience (£8000).

Mr Morrow's normal method of preparing a report was to dictate his report (not notes for it) into a dictating machine as he walked around the property. As a consequence he had no written notes. The Judge said that this practice led to a report which was 'strong on immediate detail but excessively, and I regret to have to say negligently, weak on reflective thought'.

Further the Judge held that although a departure from the recommendations contained in the RICS Practice Note 'Structural Surveys of Residential Property' was not in itself an indication of negligence there had been an inappropriate departure by the defendant in the way in which he prepared his report.

As a general rule therefore surveyors undertaking Building Surveys should take written notes on site using an A4 Clipboard or similar and these notes should be retained on the file. Recommendations made by the RICS or other professional bodies should be followed. A surveyor who dictates on site without notes or who otherwise fails to follow established professional guidance is taking a serious risk.

Merrett *v.* Babb *(2001) EGCS 20, [2001] All ER (0) 183 (Feb)*

This case went to the High Court and then, because of important implications for surveyors generally, it went to the Court of Appeal with the appeal funded by the Royal Institution of Chartered Surveyors. The case was one of professional negligence arising out of a building society valuation rather than a Building Survey. The issue in dispute was whether a surveyor employed by a firm of surveyors, not being a partner or director, was personally liable for the damages awarded if the firm for whom he had worked had gone out of business and was uninsured.

Clearly this would have very serious implications for surveyors working for small- or medium-sized firms who might be faced with a substantial claim for damages some years after the event and for which they may have no personal PI insurance cover.

The claimant purchased 18 Trelawney Road, Falmouth, Cornwall with a mortgage from the Bradford and Bingley Building Society following an inspection by the defendant who was an employee of a firm of surveyors, Clive Walker Associates. The valuation was dated June 1992 and was on the Bradford and Bingley's normal report form. The defendant valued the property for mortgage purposes at £47 500.

Applying *Smith* v. *Eric S. Bush* (a firm) and *Harris* v. *Wyre Forest District Council* (1989) 1 EGLR 169 the judge held that the defendant had owed the claimant and her mother a duty of care and that the report had failed to sufficiently notice and report on settlement cracks between an original building and an extension. The judge awarded the claimant damages of £14 500 for the defendant's overvaluation.

On appeal the defendant contended, *inter alia*, that he did not owe the claimant a duty of care because he was an employee of the firm instructed by the building society and he had not assumed personal responsibility. The Court of Appeal dismissed the appeal and held that the defendant was a salaried employee of the firm and as a professional man he must have realised that the purchaser would rely upon him to exercise proper skill and judgment and that it was reasonable and fair that the purchaser would do so. Thus although the defendant was an employee he personally owed a duty of care to the claimant.

The Royal Institution of Chartered Surveyors was not very happy about this judgment and it gave rise to a number of questions regarding the arrangements for PI insurance for employee surveyors.

Farley *v.* Skinner *(House of Lords decision reported in* The Times, *15 October 2001), [2001] All ER (0) 153 (Oct)*

This case went from the High Court to the Court of Appeal and then to the House of Lords. Mr Farley, a successful businessman, was interested in a country house about 15 miles from Gatwick Airport in Sussex and he employed Mr Skinner to undertake a survey of the property and advise him whether it would be affected by aircraft noise. Mr Skinner prepared a report on the house and in relation to the spe-cific request for advice on aircraft noise he said that he thought it 'unlikely that the property will suffer greatly from aircraft noise although some planes will inevitably cross the area, depending upon the direc-tion of the wind and the positioning of the flight paths'.

On the strength of the report Mr Farley went ahead with the purchase. On moving in he soon dis-covered that the house was affected by aircraft noise. It was close to the 'Mayfield Stack', a navigation beacon around which aircraft would be stacked up whilst awaiting a landing slot. Planes frequently flew directly over or close to the house. Mr Farley sued Mr Skinner for breach of the contractual duty of care claiming damages for diminution in value and interference with his enjoyment of the property caused by aircraft noise.

The High Court judge found the defendant negligent in his investigation of the potential for aircraft noise but dismissed the claim for damages for diminution in value on the basis that the price Mr Farley had paid was the open market value taking into account the aircraft noise. However, he awarded £10 000 for distress and inconvenience on the basis that it was a specific term of Mr Farley's contract with Mr Skinner that Mr Skinner would investigate the question of aircraft noise.

Mr Skinner appealed to the Court of Appeal and the Court of Appeal decided that Mr Farley was not entitled to his £10 000 damages on the basis that the contract was for the defendant to provide a report providing information on the property and not a contract to produce a particular pleasurable result.

Mr Farley then appealed against the Court of Appeal decision to the House of Lords which overruled the Court of Appeal and decided that the issue had been correctly decided by the judge in the High Court in the first place. The House of Lords ruled that it was not necessary for the sole purpose of the contract to be pleasure for the Claimant; it was sufficient if a major or important part of the contract was to pro-vide pleasure, relaxation or peace of mind. So Mr Farley got his £10 000 in the end and all the lawyers involved in the case no doubt somewhat larger sums.

Two important lessons for surveyors here. First, when you embark upon litigation you never know for certain how it will turn out in the end. Secondly, the terms and conditions of engagement for a Building Survey or any other type of report should limit the surveyor to his or her particular field of expertise and if a client asks for something more it is wise to decline any such requests. The client should be referred to a specialist on aircraft noise or whatever else that he or she is concerned about.

Surveyors naturally want to help their clients and it may be tempting for some surveyors to advise on all sorts of things which have nothing to do with the building such as suitability and availability of local schools (a minefield), possible effects of proposed new roads or redevelopment, aircraft noise, the neighbours (another potential disaster area) rural smells from pig farms and the like (highly seasonal) or public transport.

Generally the surveyor's PI insurance does not cover advice which is not within the surveyor's particu-lar area of expertise. So in addition to having to face a claim from an unhappy client the surveyor may find that the claim has to be faced without the benefit of insurance cover.

Some further points arising out of legal decisions

Eley *v.* King and Chasemore *(1989) 1 EGLR 181, [1989] NLUR 791, [1989] 22 EG 109*

Surveyors undertaking a Building Survey inspected inside the roof and found no evidence of water pene-tration and they inspected externally from ground level. The Court of Appeal held that the surveyors were not negligent in failing to use a long ladder to reach the roof itself for a closer inspection. The surveyors also drew attention to a tall fir tree close to the house and the attendant risk of subsidence and advised the purchaser to obtain insurance against ground movement. Here again the Court of Appeal held the survey-ors not negligent in failing to investigate further or advise the client against purchasing. The Court held that the surveyors had fulfilled their duty to the purchaser by giving advice to insure.

I have some misgivings about the decision in relation to insurance since insurance companies will generally not cover defects already existing before they came on risk (see p. 36) so a purchaser may have difficulty in practice in obtaining the necessary cover.

Hacker *v.* Thomas Deal & Co *(1991) 2 EGLR 161, [1991] 44 EG 173*

A surveyor undertaking a Building Survey did not use a torch and mirror to check for damp behind kitchen cupboards. He was held not negligent since the RICS guidance applicable at the time specifically treated the use of a mirror as a matter of individual choice.

Pfeiffer *v.* E. E. Installations *(1991) 1 EGLR 162, [1991] 22 EG 122*

By way of contrast a firm of central heating contractors testing a heating system was held negligent for not using a torch and mirror to check for cracks in a heat exchanger.

Whalley *v.* Roberts and Roberts *(1990) 1 EGLR 164, [1990] 06 EG 104*

A surveyor undertaking a mortgage valuation was not to be expected to carry and use a spirit level.

Kerridge *v.* James Abbot & Partners *(1992) 2 EGLR 162, [1992] 48 EG 111*

A surveyor is under no duty to remove stonework from a parapet wall in order to inspect a concealed part of a roof.

Bishop *v.* Watson, Watson and Scoles *(1972)*

A surveyor undertaking a visual survey of a cavity wall was held to be under no obligation to uncover those parts of the building where flashings or damp-proof courses should be found to see if they are there and if they are adequate.

Howard *v.* Horne & Sons *(1990) 16 Con LR 55, [1990] 1 EGLR 272*

A surveyor reported that electric wiring was in polyvinyl chloride cable implying that it was modern and gave no cause for concern whereas in reality the wiring was old and dangerous. The surveyor was held negligent despite conditions of engagement stating that services would only be visually inspected and would not be tested.

Heatley *v.* William H. Brown Ltd *(1992) 1 EGLR 289*

A surveyor was unable to gain access to the roof void and stated this in his report which went on to describe the house as 'in reasonable condition for its age'. He was held negligent in not advising his client to delay the purchase until access to the roof voids could be obtained and an inspection carried out.

Daisley *v.* B.S. Hall & Co *(1972) 225 Estates Gazette 1553*

A surveyor failed to report the significance of a row of poplar trees near a house and that the subsoil was a shrinkable clay.

Henley *v.* Cloak & Sons *(1991)*

A mortgage valuer reported distorted bay windows but concluded that this was due to bomb damage during the Second World War. It turned out that the defect was due to serious subsidence and that there were other observable signs of movement. The valuer was held negligent.

Allen *v.* Ellis & Co *(1990) 1 EGLR 170, [1990] 11 EG 78*

A surveyor described a garage in the curtilage of a house as built in 9-inch brick and in satisfactory condition whereas in reality it was built in clinker breeze blocks with a 50-year-old corrugated asbestos–cement sheet roof in very poor order. The surveyor was held negligent.

Marder *v.* Sautelle & Hicks *(1988) 2 EGLR 187, [1988] 41 EG 87*

A surveyor negligently failed to report that the house was built in Mundic blocks (cast from a mixture of cement and mine waste used for a time in Cornwall).

Peach *v.* Iain G. Chalmers & Co. *(1992) 2 EGLR 135, [1992] 26 EG 149*

A mortgage valuer negligently failed to report that a house was of non-standard, precast concrete panel construction (a 'Dorran' type) on which mortgage lenders do not normally lend and with limited marketability.

Hooberman *v.* Salter Rex *(a firm) (1984) 1 Con LR63, [1985] 1 EGLR 14, [1984] CILL 128, 274 Estates Gazette 151*

A flat roof used as a roof terrace was leaking and unvented leading to a serious outbreak of dry rot. The surveyor was found negligent in failing to warn of the defects including inadequate felt upstands, a lack of zinc or lead flashings and inadequate ventilation under the decking.

Gardner *v.* Marsh and Parsons *(1997) 3 All ER 871, [1997] 1 WLR 489 75 P + CR 319, [1996] 46 LS Gaz R 28, [1997] 15 EG 137, 140 Sol Jo LB 262, [1997] PNLR 362*

The surveyor failed to advise his client about over spanned floors in a five storey house converted into flats. It was held that there were signs of wallpaper drag and indications of sagging to the floors. The surveyor was held to have been negligent in failing to spot that the conversion had been defectively carried out.

Smith *v.* Eric S. Bush *(1990) 1 AC 831, [1989] 2 All ER 514*

An unsupported overhanging chimney breast in the roof space collapsed through the main bedroom ceilings about 18 months after the purchaser moved in. A mortgage valuer noted that the chimney breasts in the bedrooms had been removed but failed to check by a 'head and shoulders' inspection of the roof space whether the chimney work above had been given adequate alternative support.

Sneesby *v.* Goldings *(1995) 45 Con LR 11, [1995] 2 EGLR 102, [1995] 3 EG 136*

Another mortgage valuer found negligent for failing to report an inadequately supported chimney breast.

Bere *v.* Slades *(1989) 25 Con LR1, [1989] 2 EGLR 160, [1989] 46 EG 10 6 Const LJ 3*

A mortgage valuer found not to be negligent in failing to discover that certain walls were of non-standard and unstable construction since the defect could not have been detected by visual surface examination.

Roberts *v.* J. Hampson & Co *(1989) 2 All ER 504, [1990] 1 WLR 94, 20 HLR 615, [1988] 2 EGLR 181, [1988] NLJR 166, [1988] 3 EG 110*

The Judge considered that a mortgage valuation inspection normally takes no more than half an hour but the root of the surveyor's obligation is to take reasonable care and that if an inspection of a particular property required 2 h this is something the mortgage valuer must accept. The valuer is not normally expected to move furniture or lift carpets however if there is some specific ground for suspicion and the trail of suspicion leads behind furniture or under carpets the valuer must take reasonable steps to follow the trail until he has all the necessary information. The surveyor in this case, undertaking a mortgage valuation, was held negligent for failing to follow up clear signs of damp and thus discover a serious problem.

Lloyd *v.* Butler *(1990) 2 EGLR 155, [1990] 47 EG 56*

Another mortgage valuation case. The Judge agreed that an inspection would take on average no more than 20 to 30 min and described it as 'effectively a walking inspection by someone with a knowledgeable eye, experienced in practice, who knows where to look, to detect either trouble or the potential form of trouble'.

Nash *v.* Evans & Matta *(1988)*

A mortgage valuer who failed to detect cavity wall tie failure was held not to be negligent.

Gibbs *v.* Arnold Son and Hockley *(1989) 2 EGLR 154, [1989] 45 EG 156*

The Court agreed that a mortgage valuation inspection was limited to a 'head and shoulders' inspection of the roof space only.

Ezekiel *v.* McDada *(1994) 2 EGLR 107, [1995] 47 EG 150, 10 Const LJ 122*

A mortgage valuer found to be negligent for failing to report defects which would have been apparent from a 'head and shoulders' inspection.

Sneesby *v.* Goldings *(1995) 45 Con LR 11, [1995] 2 EGLR 102, [1995] 36 EG 136*

Surveyor carrying out a mortgage valuation held negligent for failing to check that a chimney breast over a kitchen which had been removed was adequately supported, this being a matter which would have been apparent from looking inside the kitchen cupboards or cooker hood (something which he would not normally be expected to do on a limited inspection of this type).

Hipkins *v.* Jack Cotton Partnership *(1989) 2 EGLR 157, [1989] 45 EG 163*

A surveyor undertaking a Building Survey must also follow the trail. In this case cracks in the rendering and a sloping site should have raised serious questions in the mind of a reasonably prudent surveyor as to the adequacy of the foundations.

Hingorani *v.* Blower *(1976) 1 EGLR 104, 238 Estates Gazette 88, [1976] EGD 618*

A property had been recently subject to extensive redecoration and modernisation, and this itself should give a surveyor reason to check carefully that the vendor had not sought to conceal serious defects. In this case subsidence which required underpinning.

Lloyd *v.* Butler *(1990) 2 EGLR 155, [1990] 47 EG 56*

Further clarification from the Judge on the question of 'following the trail'. A surveyor does not have to follow up every trail to discover whether there is trouble or the risk of trouble. But where there is evidence of actual or potential trouble the surveyor must report in such a manner as to alert the client and anyone else entitled to rely on the report to the risk.

Unfair Contract Terms Act 1977

This applies to England and Wales, but not to Scotland, and limits the extent to which a contract may attempt to exclude liability. Before a surveyor attempts to limit liability he or she should confirm, by taking professional advice if necessary, that it is possible to do so having regard to the provisions of this Act. A surveyor will also have to consider if a limitation of liability is ethically appropriate having regard to the circumstances.

In Section 2 of the Act provisions are made in respect of negligence liability. The Act states:

'1. A person (in the course of business) cannot by reference to any contract term or to any notice given to persons generally or to particular persons exclude or restrict his liability for death or personal injury resulting from negligence.'

'2. In the case of other loss or damage, a person cannot so exclude or restrict his liability for negligence except in so far as the term or notice satisfies the requirement of reasonableness.'

In these contexts the term 'business' includes a profession.

In Section 3 the Act provides certain rules in respect of liability arising in contract, and states:

'1. This section applies as between contracting parties where one of them deals as consumer or on the other's written standard terms of business.'

'2. As against that party, the other cannot by reference to any contract term (a) where himself in breach of contract, exclude or restrict any liability of his in respect of the breach; or (b) claim to be entitled (i) to render a contractual performance substantially different from that which was reasonably expected from him; or (ii) in respect of the whole or any part of his contractual obligation, to render no performance at all, except in so far as (in any of the cases mentioned in the above sub-section) the contract term satisfies the requirement of reasonableness.'

The 'requirement of reasonableness' means that the contract term should be fair and reasonable having regard to all the circumstances of which the parties were aware, or ought to have been aware, at the time the contract was entered into.

A distinction is made between those cases where a surveyor is acting for a client as consumer and those cases where the client is not a consumer, 'consumer' for this purpose meaning an ordinary member of the public making use of the surveyor's services in his private capacity in connection with his personal affairs. A client who is him/herself in business, or a company client, will not be a consumer for this purpose and as between such non-consumers and their surveyors the Act allows special bargains to be struck which can exclude any need to consider if the test of reasonableness needs to be applied.

A member of the public buying a house is a consumer for this purpose. Whatever are the terms agreed between the surveyor and such a client they must pass the test of reasonableness. The normally accepted and traditional practices of a profession, in respect of the nature of the inspection and the matters covered in the report, would be a reasonable basis for a contract between the parties. A substantial departure from normal or traditional surveying practices might be unable to pass a test of reasonableness.

In so far as a surveyor follows traditional and normal practices, and the recommendations of his professional institution, it would seem most unlikely that this would be held to be unreasonable.

Supply of Goods and Services Act 1982

This Act, which came into force on 4 July 1983, codifies the common law position as already described in this chapter, and covers any contract under which a person agrees to carry out a service. The Act provides implied terms which are deemed to be incorporated into every contract for the supply of a service in that the supplier (the surveyor) will exercise reasonable care and skill, that he will carry out the service within a reasonable time and that the charge made will be a reasonable one (terms as to time and charges apply only when these matters are not covered by the contract itself). Only if the contract fixes neither the fee nor the manner of its assessment (i.e. an hourly rate for professional time) may a client challenge a charge made using the Act's provisions.

It is felt that the Act will not make any significant change in the legal liability for negligence in surveying buildings, merely that it has codified existing law by statute.

Complaints handling procedures and PI insurance

All members of the Royal Institution of Chartered Surveyors undertaking Building Surveys and related activities are required to have in place a complaints handling procedure and to be covered by PI insurance.

PI insurance is provided by insurers on what is termed a 'claims made' basis. The policy is written so that the cover is provided in respect of claims made during the currency of the policy irrespective of when the alleged negligent act took place. So the insurers of a practice today will be providing cover in respect of claims arising out of work undertaken some years ago, perhaps as long as 15 years or more years ago. Policies of this type are known as 'fully retroactive' and will cover the insured for claims arising out of the practice's work since the establishment of the business.

So if you change insurers the old insurers will no longer have any liability in respect of claims about which they were not already notified, and the new insurers will be taking on the potential liability for all the work previously done by the practice. Not surprisingly the proposal form which needs to be completed for the new insurance policy will require lots of information regarding the past activities of the firm and any knowledge there may be regarding abnormal risks or potential claims.

It goes without saying that it is a principle of insurance law and practice that there must be full disclosure by the insured of any matters which could have a bearing on the insurer's risk and if it is found later that the insured was less than frank when completing the proposal form the insurers can repudiate liability.

Costs and expenses in defending a claim are included under the policy and are paid in addition to the indemnity limit. The RICS requirements for PI cover include provision that indemnity must be on an 'each and every claim' basis. So instead of providing cover of (say) £1 million over a 12-month period the underwriters will provide cover of £1 million in respect of each and every claim.

The insurers will always insist that the policy is subject to an 'excess'. That is to say the insured will bear the first part of any claim up to an agreed limit. So there may be cover for £1 million in respect of each and every claim but subject to an excess of, say, £10 000 in each case. Insurers hope that the imposition of an excess will ensure a degree of caution in the insured's activities. The insured must then contest or settle the small claims up to £10 000 without recourse to the cover provided by the policy. It is always necessary, however, for the insurers to be notified of any claim or any incident which could give rise to a claim, however small, and they should be kept advised of the progress of the matter.

There is always the risk that a small claim, which appears to be covered by the excess, may suddenly degenerate into a much larger claim with implications for the insurers.

PI premiums have rocketed over recent years and cover is now very expensive so small surveying practices, in particular, are having to pay a significant proportion of their fee income for insurance. One way to reduce the cost is to agree to a larger excess. It also helps if you have a good track record for claims, with few or no complaints that need to be notified.

20
Dilapidations

The rising damp is cured now – the dry rot soaks it up.

Evil landlord to gullible tenant

In addition to building surveys for intending purchasers and tenants, surveyors will also be asked from time to time to carry out surveys in connection with dilapidation claims, acting for either the landlord or the tenant.

Before proceeding it is essential that the surveyor understands the law involved because he or she will need to be able to read and interpret the lease and consider the application of certain statutes which provide some protection for tenants against claims in many situations. Although most claims are by landlords against tenants it is also possible for a tenant to have a right to claim against a landlord for breaches of repairing covenants and surveyors may occasionally be asked by a tenant to act in such a matter. The law referred to in this chapter is that applicable to England and Wales.

Claims by tenants against landlords of residential property sometimes feature in the law reports because tenants are generally legally aided or supported by housing action groups of one kind or another and many of the claims by tenants in recent years have been against local authority landlords. For surveyors in private practice the bulk of dilapidations work will involve commercial premises and the majority of claims will be by landlords.

The first thing that ought to be said is that if buildings are in poor repair and if the tenants are responsible they should do the work. There is nothing to be gained in engaging in a prolonged dispute about repairs which clearly need doing. In fact many good buildings are allowed to deteriorate due to poor maintenance and ineffective property management and if managing agents were more diligent the country's building stock would be in a better state.

This said there will undoubtedly be cases from time to time where a tenant wishes to dispute a landlord's Schedule of Dilapidations and in England and Wales there are nine principal grounds for objection which can normally be used. When acting for landlords or tenants it is worthwhile checking through these nine important matters to confirm whether, and to what extent, they may apply to the case under review.

The nine grounds for objection are as follows:

1. Defective notices

Material errors may result in a landlord's schedule being unenforceable. This will apply particularly if a landlord is serving notice under Section 146 of the Law of Property Act 1925 and claiming damages or forfeiture. These notices have to be served in a particular form and the schedule must specify precisely what work is required. A vague schedule is not acceptable to the courts.

If the lease was originally granted for a term of more than 7 years and there are more than 3 years to run the notice must include certain additional information required by the Leasehold Property (Repairs) Act 1938 – an act intended to protect tenants.

These formal notices are best served by a solicitor using the schedule prepared by the surveyor. I do not recommend that surveyors serve notice of disrepair themselves claiming damages or forfeiture having regard to the legal processes which may follow if the claim is disputed.

Schedules are usually described as being either Interim Schedules or Terminal Schedules. These terms have no particular legal significance. In law a schedule may be served at any time whilst the lease is running (subject to the tenant's right to apply for relief under the Leasehold Property (Repairs) Act) or for up to 12 years after the expiry of the lease if it is by deed or 6 years after if it is not by deed (Limitation Act 1939 Section 2). Any financial settlement of a terminal schedule will, however, normally be in full and final settlement of all claims so it is more important with a terminal schedule to ensure that nothing is missed. If an interim schedule is served and something is missed the landlord can usually serve a fresh schedule at a later stage.

2. Repairs may be outside the demised property

Before serving any notice on behalf of the landlord, or accepting it on behalf of the tenants, a check is advised of the description of the demised property in the lease, and the lease plan. Clearly if the repair required, or the compensation demanded, relates to matters outside the demise or outside of any common parts referred to in the lease, the tenant is not liable.

For example, in *Hatfield* v. *Moss* [1988] 40 EG 112 the Court of Appeal had to consider whether the roof space on the top floor of a building divided into six flats was included in the demise of the top floor flat. The lease did not mention the roof space and a plan attached to the lease showed a roof space outside the line enclosing the demised property. The defendant had converted the roof space into a playroom and storeroom. See also *Straudley Investments Ltd* v. *Barpress Ltd* [1987] 282 EG 1224 where the issue turned on the rights to construct a fire escape and ventilation vents on and against the roof of the demised property.

In *Hatfield* the Court of Appeal held that a plan which is for identification only cannot control the parcels clause in the lease which describes the extent of the demise, however if the parcels clause is unclear then it was appropriate to look at the plan for guidance regarding what may be included or excluded within the demised property. Lease plans are normally described as being attached for identification purposes only, and not all leases have plans. So it is the written description which one must consider first and if this is clear then whatever is shown on the plan does not change the position. Only if the written description is unclear can one resort to using the plan and any surrounding circumstances to settle the issue.

3. The lease may not be on full-repairing terms

A careful reading of the lease may show that the Repairing Covenants do not include the repair required: in other words the lease may be something less than a full-repairing lease.

Covenants that refer to 'Good and Substantial Repair' will normally be effective as full-repairing covenants. Covenants that refer to 'Good and Tenantable Repair' require a standard of repair suitable for a tenant carrying on the particular kinds of businesses for which the property is leased which, in the case of some types of industrial or retail premises, may not be as high as 'Substantial' repair. Some covenants refer only to a tenant's liability for interior or exterior decorations and may be silent on the question of liability for major repairs, or may require that the landlord repairs the 'exterior and structure'.

If in doubt it is best to refer to the solicitor for an interpretation of any covenants that appear doubtful. If a schedule is served on behalf of the landlord it could rebound to his disadvantage if it includes matters which are found, on examination, to be landlord's repairs.

4.   Section 18 (1) of the Landlord and Tenant Act 1927 may apply

Section 18 (1) of the Landlord and Tenant Act 1927 provides that a ceiling is set over any landlord's claim for damages which cannot exceed the loss in the value of his interest in the property (diminution in the value of his reversion). For example, if the landlord can be shown to be proposing to alter or demolish a property its state of repair may be irrelevant and the tenant may happily allow it to deteriorate, notwithstanding that he may have a full-repairing lease.

The section reads as follows: 'Damages for the breach of a covenant or agreement to keep or put premises in repair during the currency of a lease or to leave or put premises in repair at the termination of a lease, whether such agreement is expressed or implied and whether general or specific, shall in no case exceed the amount (if any) by which the value of the reversion (whether immediate or not) in the premises is diminished owing to the breach of such covenant or agreement as aforesaid; and in particular no damages shall be recovered for a breach of any such covenant or agreement to leave or put premises in repair at the termination of a lease, if it be shown that the premises, in whatever state of repair they might be, would at or shortly after the termination of the tenancy have been pulled down, or such structural alterations made therein as would render valueless the repairs covered by the covenant or agreement.'

With some types of property the state of repair of particular items may be unimportant in terms of letting value. For example, shop premises are invariably refitted by the incoming tenant, often at great expense to suit a particular corporate image, so the state of the interior and shop front left by the outgoing tenant may not be important.

In *Mather* v. *Barclays Bank Plc* [1987] 2 EGLR 254 the landlord claimed against an outgoing tenant for substantial dilapidations, however new tenants were found (a building society) who undertook not only the repairs but also substantial improvements. The investment value of the landlord's reversion following the letting to the new tenant, capitalising the rent at a 7% yield, produced a valuation well above that for the property repaired but unimproved. The court held that the landlord's claim failed under the first limb of Section 18 (1) of the Landlord and Tenant Act 1927 as the value of the reversion had not – as it turned out – been diminished by the breach of covenant.

Outgoing tenants would be wise to delay settlement of a claim for as long as possible to see what happens to the property on reletting.

Another case dealing with the application of Section 18 of the 1927 Landlord and Tenant Act was *Mason* v. *Totalfinaelf UK Ltd* (2003) EWHC 1604 (Ch.) (2003) 30 EG 145 (CS). This concerned a terminal dilapidations claim made in respect of premises in Crawley trading as petrol station, shop and garage which had been leased to Total in 1964 and where, on termination of the tenancy, a dispute had arisen as to the extent of the outgoing tenant's liability for dilapidations.

The issues were the extent to which the tenants had breached their repairing liabilities, the reasonable cost of remedying the disrepair for which they were liable and the extent to which the breaches of covenant had diminished the value of the freehold. It was accepted that the claimants could only recover costs of repairs in so far as they did not exceed the diminution in the value of the reversion.

The defendants had covenanted to repair and keep in good condition the demised premises to the satisfaction of the landlord's surveyor. During the trial the judge went through a long list of disputed repairs. On a number of occasions the judge concluded that the landlord's case had not been made out because specialist reports recommended by his surveyor had never been obtained. In respect of the two largest elements of the claim, replacement of underground petrol tanks and dealing with ground contamination, the judge ruled against the landlord because there was no evidence that a problem had yet occurred. The specialist report on the tanks did no more than indicate that they were now of an age when they could be expected to need replacement but there was no evidence of actual disrepair.

In the end the judge concluded that the reasonable cost of the works for which the tenant was legally liable was £135 000. He then went on to consider how this is related to the trading potential of the three aspects of the business operating from the premises; that is the petrol station, the shop and the motor trade element. All were diminished in value and he determined the total to amount to £73 500. This was much less than the cost of the repairs and accordingly the landlord's damages were capped at the lower figure.

A number of other interesting issues were raised. The judge remarked that the tenant's surveyor had not understood his duties as an expert witness; rather he had seen his task as helping his clients to keep their liability for dilapidations to a minimum and this had caused the court to have less confidence in his assessment than would otherwise be the case. (For more information regarding the duties of surveyors acting as expert witnesses see Chapter 22.)

Also the lease had said that the repairs had to be carried out to the satisfaction of the landlord's surveyor; the judge held that this provision did affect both the content and the manner in which the work should be undertaken but the landlord's surveyor cannot require the tenant to do anything which falls outside the covenant and this provision did not give the landlord's surveyor *carte blanche*. The landlord's surveyor can however require more extensive work than those being proposed by the tenant's surveyor provided that his decision is reasonable.

Readers interested in the workings of Section 18 (1) would be advised to read the judgement in full.

5. Internal decorations may be excluded

It may come as a surprise to many tenants to know that a landlord cannot automatically require internal decorations to be carried out during the term of a lease and an interim schedule which includes internal decorations can be challenged by an application to the court for relief. (A terminal schedule at the end of the lease may quite properly include internal decorations however.) This is provided for in Section 147 of the 1925 Law of Property Act. Note however that decorations which are required to prevent deterioration to building components or ensure compliance with covenants to repair or ensure cleanliness must still be undertaken.

The section provides as follows:

1. After a notice is served on a lessee relating to the internal decorative repairs to a house or other building, he may apply to the court for relief, and if, having regard to all the circumstances of the case (including in particular the length of the lessee's term or interest remaining unexpired), the court is satisfied that the notice is unreasonable, it may, by order, wholly or partially relieve the lessee from liability for such repairs.
2. This section does not apply:
    (i)   Where the liability arises under an express covenant or agreement to put the property in a decorative state of repair and the covenant or agreement has never been performed;
    (ii)   to any matter necessary or proper –
           (a) for putting or keeping the property in a sanitary condition, or
           (b) for the maintenance or preservation of the structure;
    (iii) to any statutory liability to keep a house in all respects reasonably fit for human habitation;
    (iv)   to any covenant or stipulation to yield up the house or other building in a specified state of repair at the end of the term.

6. Tenant may not be liable to renew, as opposed to repair

This is a more arguable area since the protection provided for the tenant is not a statutory protection but the protection of the common law and subject to legal precedent. In general the courts have held as a matter of principle that a tenant is not liable to hand back anything which is improved to the point where it is wholly different from that originally demised. Repair will include some replacement of parts in many cases but should not extend to the point where it constitutes renewal of the whole. (A lease may, of course, provide for renewal rather than repair but a sensible tenant should avoid any such commitment unless the implications are fully understood and acceptable.)

The extent of a repairing covenant can have implications in other areas of Landlord and Tenant Law such as rent review. In *Norwich Union Life Assurance Society* v. *British Railways Board* [1987] 283 EG 846 a repairing covenant in a 150-year lease with provisions of rent reviews at 21-year intervals required the tenant to 'keep the demised premises in good and substantial repair and condition and when necessary to rebuild, reconstruct or replace the same'. An arbitrator dealing with a rent review dispute considered that this placed on the tenant a more onerous obligation than the usual covenant to keep the premises in good and substantial repair and he accordingly made a downward adjustment of 27.5% in the rent to reflect the difference.

The court upheld the arbitrator's view and the appeal against his award was dismissed. The judgment recorded that the language of the repairing covenant included two separate obligations; first, that of repair and secondly, that of rebuilding, reconstructing or replacing the entire premises.

In a normal case what actually constitutes repair or renewal must depend on the circumstances. Underpinning of part of a building has been held to be a repair notwithstanding that it is also an improvement to remedy an inherent defect (see *Rich Investments Ltd* v. *Camgate Litho Ltd* [1988] EGCS 132).

Eventually a roof covering needs renewal but in *Murray* v. *Birmingham City Council* [1987] 283 EG 962 the court held that the roof of a terraced house built in 1908 had not yet reached the stage where complete renewal of the roof was the only option and piecemeal repairs were still a valid option for the landlord. In this case the tenant on a weekly tenancy wanted a new roof and the local authority landlords wanted to carry on undertaking patching repairs to the existing roof.

7.   The Leasehold Property (Repairs) Act 1938 may apply

In any case where a lease was granted for a term of more than 7 years and there are more than 3 years to run the landlord may not proceed with service of notice of disrepair and Schedule of Dilapidations without the leave of the court.

A preliminary court hearing is required and the burden of proof is on the landlord to show that the repairs are required on one of five grounds laid down in the Act, the most important of which is that the repair is required to prevent substantial diminution in the value of the landlord's interest. As Lord Denning MR said (*Sidnell* v. *Wilson* 1966): 'That Act was passed shortly before the war because of a great mischief prevalent at that time. Unscrupulous people used to buy up the reversion of leases, and then bring pressure to bear on tenants by an exaggerated list of dilapidations.'

At the preliminary court hearing surveyors will often be called to give evidence on the matter of the state of repair of the premises and the significance of dilapidations as they may affect the value.

The five grounds are as follows:

(a)   That the immediate remedying of the breach in question is requisite for preventing substantial diminution in the value of the reversion, or that the value thereof has been substantially diminished by the breach.

(b)   That the immediate remedying of the breach is required for giving effect in relation to the premises to the purposes of any enactment, or of any byelaw or other provision having an effect under an enactment, or for giving effect to any order of a court or requirement of any authority under any enactment or any such byelaw or other provision as aforesaid.

(c)   In a case in which the tenant is not in occupation of the whole of the premises as respects which the covenant or agreement is proposed to be enforced, that the immediate remedying of the breach is required in the interests of the occupier of those premises or part thereof.

(d)   That the breach can be immediately remedied at an expense that is relatively small in comparison with the much greater expense that would probably be occasioned by postponement of the necessary work.

(e)   Special circumstances which in the opinion of the court render it just and equitable that leave should be given.

Note: These paragraphs should be read as if 'or' came between each of them so that the landlord need only satisfy the court on one ground (see *Phillips* v. *Price* [1959] Ch. 181).

For practical purposes the effect of this Act is to prevent a landlord serving a notice of disrepair and schedule until the last 3 years of the term except in cases of fairly extreme disrepair. For managing agents perhaps it would be sensible to make a diary note to inspect all properties where the original leases were for 7 years or more at the time when they have 3 years to run. If there are going to be arguments about dilapidations there is then plenty of time to inspect, serve notice and follow the matter up before the lease actually expires.

8.   A schedule of condition may exist

When a lease is granted it is often a good idea for landlord and tenant to agree a schedule of condition, describing the property fully, as evidence of its state of repair and then have copies attached to the lease and counterpart lease. This is particularly important from the tenant's point of view if the property is old and in poor repair. The lease may then provide that, notwithstanding the provisions of any repairing covenants, the tenant will not be liable to hand the property back to the landlord on termination in any better condition than is evidenced by the schedule of condition.

Surveyors acting for tenants in receipt of dilapidations claims would be advised to check the lease to see if there is any reference to a schedule of condition and if so to locate a copy of the schedule.

The schedule normally consists of an item-by-item description of the property and its various components, frequently with photographs.

9.   The landlord may be liable for implied repairing covenants

Sections 11–16 of the Landlord and Tenant Act 1985 (formerly Sections 32–33 of the Housing Act 1961) provide that when a residential tenancy is granted for less than 7 years the landlord always remains liable for certain matters. It is not possible to contract out of this liability, except by granting a residential lease of more than 7 years.

The implied covenants are as follows:

(a)   to keep in repair the structure and exterior of the dwelling-house (including drains, gutters and external pipes);
(b)   to keep in repair and proper working order the installations in the dwelling-house for the supply of water, gas and electricity and for sanitation (including basins, sinks, baths and sanitary conveniences but not other fixtures, fittings and appliances for making use of the supply of water, gas or electricity);
(c)   to keep in repair and proper working order the installations in the dwelling-house for space heating and heating water.

It should be noted that Section 11 of the 1985 Act is amended by Section 116 of the 1988 Housing Act in respect of leases granted on or after 15 January 1989. Under these provisions the implied covenants extend to 'any part of the building in which the lessor has an estate or interest' if the 'Disrepair . . . is such as to affect the lessee's enjoyment of the dwelling-house or of any common parts, as defined in Section 60 (1) of the Landlord and Tenant Act 1987, which the lessee, as such, is entitled to use'. This extension of the lessor's liability is intended to cover situations which may well arise in blocks of flats or buildings in multiple occupation.

Taken together these provisions are intended to ensure that tenants in short term residential lettings can insist on certain basic minimal standards of maintenance and the provision of basic sanitation including hot water and central heating where these exist at the commencement of the letting. These provisions do not apply to lettings of commercial premises.

The inspection

The foregoing is a very condensed summary of the legal position as it applies to notices of disrepair and the negotiations which may take place between surveyors acting for landlords and tenants. Most claims are settled amicably by negotiation against this legal background.

Since this is not intended as a legal text book I have not attempted to cover the legal aspects in any detail and readers interested in delving more deeply into the complexity of dilapidations claims will be able to find much of interest in the various legal text books.

From the surveyor's point of view these matters generally arise initially with a request from a landlord to inspect a property and advise whether a schedule is required, or from a tenant who is in receipt of an unwanted schedule. Often an informal list of repairs is prepared first and sent, with an initial letter, to the tenant pointing out the various matters that need attention. This informal letter may be followed by a formal notice of disrepair and schedule later if the repairs are not attended to.

The inspection procedure will be similar to that required for a building survey with a need for particular thoroughness if a terminal schedule is involved since once a full and final settlement is reached the landlord will not be able to claim for anything which may have been missed (unless, that is, he decides to claim against his hapless surveyor!).

The notes will be taken in tabular form covering all elements of the construction internally and externally and listing all work needed to comply with the repairing covenants.

If a case is likely to go before the court then it is normal for the schedule to be prepared in the form of a 'Scott Schedule' with columns for each item, notes from any schedule of condition, details of the alleged dilapidations, the landlord's claim, the tenant's response, the plaintiff's observation, the defendant's observation and a final column for the judge's notes.

During any hearing the judge may then move briskly through the schedule dealing with each item in turn and noting his award of damage (if any).

If the tenant is unwilling or unable to do the work the claim will be expressed financially and it will be necessary to cost out each item, so some measurements will generally need to be taken during the inspection in order to calculate the cost of some repairs, expressed as a price per square metre or the equivalent.

In many cases Section 18 (1) of the 1927 Landlord and Tenant Act will be invoked by the tenant, either to repudiate the claim altogether or to justify a reduction in the total claim. If it is clear that the landlord proposes to demolish or reconstruct the property in such a way that its condition is immaterial to its value to him then any claim will be unsupportable and there would seem to be little point in spending valuable time preparing a schedule in the first place.

If the tenant claims that the landlord has such intentions, but the landlord denies it – or claims to have had a change of mind – it would be sensible for the solicitors to resolve this preliminary point before instructing surveyors.

Even if the landlord has no such proposals there may be arguments about diminution in value. In such cases it is best to complete the schedule first and calculate the cost of repairs; then to look at the total arrived at in relation to the value of the property as it is and as it would be with the repairs done.

In some cases there will be no difference – the cost of repair will be the same thing as the difference in value. In other cases some repairs will be judged unimportant in relation to the value of the property.

Calculation of the amount by which the reversion has been reduced in value is undertaken on a simple 'before and after' basis taking the value of the landlord's interest (which may not necessarily be freehold) as it stands and then as it would be had no breaches occurred.

The survey for dilapidations purposes should therefore also include measurements and other details of the property because the open market value of the landlord's interest will have to be assessed. With retail premises, for example, the location from a trading point of view will be very important and probably of more significance to its value than matters such as shop fittings which have a traditionally short life. Most incoming tenants of retail units will budget for refitting the interior anyway and the actual condition of the interior may well be irrelevant to letting value (and hence the value of the landlord's reversion).

It is worth noting at this stage that the statutory protection provided for tenants does not apply the other way. If a landlord has a repairing liability and is in breach then the tenant may sue for damages. The tenant's claim will succeed or fail simply by the application of the law of contract to the particular facts as evidenced by the surveyors to each side as expert witnesses on the issues of liability, cost of repairs and incidental losses which the tenant may have suffered (e.g. from leaking roofs).

Statutory dilapidations

In addition to the liability to repair arising under the law of contract and the Housing Act there is a variety of statutes which may make property owners, occupiers or building contractors liable in various circumstances for repairs or other works to buildings in order to comply with the law. Some of the more important statutes are as follows:

Fire Precautions Act 1971 (Sections 1 and 2)

This designates premises for which fire certificates are required. The condition of fire escapes, alarm and sprinkler systems and the means of preventing spread of fire are all important features in any dilapidations survey. From 1 April 1989 all factory, shop, office or railway premises where more than 20 people work, or where more than 10 people work at any one time on a floor other than the ground floor, have required a fire certificate. Even if a fire certificate is not required, because fewer than the stipulated number of people work there, it is still necessary to provide the means of escape and the means of fighting fire. The local fire service can advise individual property owners on the action needed to comply.

Defective Premises Act 1972

This places a duty on a builder to build properly (Section 1) – a duty which is not discharged by any subsequent disposal of the building (Section 3).

Offices, Shops and Railway Premises Act 1963

Specifies minimum requirements for space, temperature, ventilation, lighting and sanitary conveniences. Note especially the requirements for adequate lavatory and washing facilities, including hot water, for both sexes.

Building Regulations

These include provisions for enforcement in respect of new buildings or building alterations. During a dilapidations survey it is necessary to look out for unauthorised alterations which may have been undertaken without either landlords' or local authority consent.

Local Acts

At one time there were hundreds of local Acts of Parliament which applied to particular boroughs, cities or counties. Most of these have been repealed but some remain in force in mainly urban areas affecting such diverse matters as access for fire fighting, storage of inflammable liquids, temporary structures, dust from buildings and the fixing of signs. If in doubt it is wise to check with the building control department of the local authority whether any local Act applies.

London Building Acts (Amendment) Act 1938

Originally a local Act covering the special code for party wall procedures within inner London, intended to facilitate the development of urban sites where a need may arise to build into or on existing party walls, or repair party walls. The Act provides a procedure for resolving disputes. The provisions were found to work well in inner London and were extended in 1996 to apply to the whole of England and Wales, under the provisions of the Party Walls (etc.) Act 1996.

Factories Acts

Certain minimum standards are provided for similar to those that apply under the Offices, Shops and Railway Premises Act.

Civil Procedure Rules

A schedule of dilapidations is an allegation of breach of covenant and if the necessary notices have been correctly served may result in court proceedings which are subject to the Civil Procedure Rules introduced in England and Wales in 1999. These rules incorporate certain protocols intended to streamline the process and the claimant in such an action is required at the outset to provide full details of the claim supported by an expert's report.

A landlord's surveyor who prepared a schedule of dilapidations may not be sufficiently independent to provide the necessary expert report should the matter go to court, so a landlord may have to employ one surveyor to prepare the schedule as a basis for the claim and a second surveyor to act as an expert witness in any court proceedings. The surveyor preparing the initial schedule is traditionally partisan, acting for his or her client, the landlord, whilst the expert witness has a primary duty to assist the court. The court has powers to impose strict penalties on claimants who inflate their claims or mislead respondents over the nature or extent of the claim.

Statements of truth have to be signed by the parties and a surveyor acting as expert witness in dilapidations matters is precluded from putting forward an inflated or unrealistic claim or a claim which fails to take into account the reliefs available to the tenant, for example by virtue of Section 18 (1) of the 1927 Act.

Surveyors who find themselves advising clients in these circumstances should ensure that they are fully familiar with the Civil Procedure Rules which are intended to facilitate the early settlement of disputes out of court and a more rapid conclusion of court proceedings should they arise (see Chapter 22). Surveyors should be familiar with the Protocols laid down for the conduct of dilapidations claims.

Conclusion

When dealing with dilapidations claims it is wise to keep in mind the importance of maintaining good relations between landlords and sitting tenants – they need one another. The tenant needs the premises – the landlord needs the income. Also the fact that dilapidations issues often arise at the same time as other negotiations for lease renewal, assignment, change-of-use or rent review. Frequently dilapidations matters are settled as part of an on-going property management strategy. Careful study of the law involved in all these areas is recommended because a dilapidations claim against a sitting tenant is rarely a matter that can be considered entirely in isolation.

21

Conservation and the surveyor

Another giant glass stump better suited to downtown Chicago than the City of London – a monstrous carbuncle on the face of a much loved and elegant friend.
HRH The Prince of Wales referring to proposed developments adjoining St Pauls, at the RIBA
150th Anniversary Banquet in 1984

There is now a welcome increase in demand for building surveys undertaken not for prospective buyers, mortgage lenders, landlords or tenants, but those undertaken for existing property owners needing advice on the conservation of their property. Conservation is now a major national and world issue. We must avoid the wasteful use of the world's resources including the fine buildings passed down to us by previous generations. We must be prudent and economical in our use of energy and building materials. I make no apologies for devoting a whole chapter to this issue.

Conservation principles

Surveyors can assist conservation in its widest sense if they aim to further the following objectives when undertaking their day-to-day work:

1. To design and construct buildings with a long lifespan requiring minimal maintenance and capable of adoption to changing needs and the requirements of occupiers.
2. Avoid the wasteful use of materials and energy in the construction process and subsequent life cycle maintenance.
3. To design buildings and modify buildings so as to make them easily repairable with convenient means of access to structural components and services.
4. To plan the maintenance of existing buildings over their life cycle and avoid damage by neglect.
5. To use sympathetic materials and techniques when maintaining existing historic buildings.

Practical conservation techniques are applied in the main to buildings which are of historic or aesthetic importance including most ecclesiastical structures, listed buildings of all types, tithe barns, country houses and even art deco and post-war property where the design or construction is of special interest. Conservation is increasingly important now with the growth of Conservation Areas designated under planning legislation, and Georgian, Victorian and Edwardian terraces are enjoying renewed popularity in many of our towns and cities.

As conservation generally requires a greater degree of maintenance of a continuing nature than would be required for a building of less importance, a great deal of thought and planning should have been employed in ensuring the most effective use of the physical and financial resources available. If the necessary finances and technical expertise have not been available in the past then the surveyor must be extremely wary and meticulous in carrying out his or her inspection.

Preparations for a survey

The agencies causing deterioration and subsequent decay in an older building can generally be attributed to the weather, vandalism and misuse, the forces of gravity, chemical attack and biological attack from fungus and insects. People are often responsible for much of the damage.

Before carrying out an inspection, and prior to preparing a building survey, one must therefore consider the original design, its structural advantages and weaknesses, undue wear or damage caused by human agencies and the natural elements and their impact. Sadly unattended older buildings require special attention and security since they were not designed or constructed to resist crowbar or aerosol.

Thankfully in this country we are fortunate in the scarcity of natural disasters such as earthquake or flood and no special precautions are needed against such eventualities but often too scant regard is paid to the effects of gale force winds and snow against which some precautions should have been taken by those responsible for the upkeep of the building.

Before dealing in detail with any of these matters, it must be remembered that during the process of any conservation work to an important building there should have been no interference with any historical evidence which may be in the building or on the site. Before commencing such work the owners should have taken particular care in compiling a fully documented record to include materials and designs used in the construction. Any such information already available should be closely studied and logged by the surveyor before he, or she, begins the survey.

Before commencing any work to a 'special' building, preparing specifications or seeking tenders a full condition survey should therefore be made and a comprehensive report on the structure prepared. This will then indicate how the materials and methods of construction used in the structure have performed in the past and facilitate an evaluation of the likely future performance of that building.

From such a report, in consultation with the owners, the 'degree of intervention' required to carry out the conservation work may be assessed, taking one of the following listed in rising order of cost and complexity:

1. Simple prevention of further deterioration.
2. Preservation in existing condition.
3. Structural consolidation.
4. Restoration of structure, fittings and finishes.
5. Rehabilitation including possible improvement items.
6. Reproduction – imitative or by substitution.
7. Reconstruction in whole or in part.

Before starting the survey it is useful for the surveyor to know in broad terms what level of expenditure and degree of intervention the owner has in mind (most owners will naturally want a Rolls-Royce job for a Morris Minor price). At this point it is also important to establish whether the works, or any part of the works, are to be grant aided from charitable or public funds, since the standards and requirements of the organisation providing the funding will have to be met.

Conservation work generally demands the services of highly skilled and experienced craftsmen (and craftswomen) – stonemasons, for example – who have a long lifetime of experience of traditional methods and materials. In many cases good stonemasons, carpenters, joiners and even plumbers must be artists just as much as tradesmen. The choice of labour for this type of work is particularly vital.

Obviously, it is important that the surveyor does not recommend solutions which are impractical because the skilled trades needed are simply not available in the locality.

To summarise these preliminaries which may be possible for the surveyor to undertake prior to the condition survey:

1. Assess environmental, human and natural agencies causing deterioration.
2. Obtain available documentation including any historical evidence.
3. Establish the degree of intervention the owner has in mind and resources available.
4. Consider the type and availability of the craftsmanship needed.

Techniques, structures and materials

Many older structures to which conservation techniques are applied are less sophisticated and more massive than modern structures which offer equivalent accommodation. Now, in the age of the computer and with the Building Regulations and Codes of Practice offering satisfactory minimum standards, buildings tend to be 'built-down' to these calculated minimum standards and so modern construction often appears to be surprisingly fragile although in fact generally adequate for the forces involved.

The reasons for the more massive material content of the older buildings are that weaker materials were used and 'rule of thumb' was the order of the day, the tendency being to ensure stability by over-designing with extravagant use of material resources.

Another reason for the provision of strength by mass and sheer size was the intended permanence of structures. Short life public buildings were probably unknown in Norman times and during the Middle Ages. Even the sophisticated Romans and Greeks built with permanence in mind and we can still benefit from the results. Compare, for example, Wells and Canterbury Cathedrals with the Roman Catholic Cathedrals of Liverpool and Bristol; which are the more massive and which are likely to have the greater permanence? But in mitigation of modern design, which are likely to be the more maintenance-intensive, original construction defects apart?

The first thoughts of a surveyor on considering the performance of an older building would be directed towards the loadings on the different parts of the structure and, in particular, the foundations. Modern road traffic – increasing in both volume and weight – uses roads which were never designed for the proper transmission of the consequent loadings and the ground vibrations thus caused by such traffic can cause foundation displacement with resultant movements in the upper structure.

Rapid and irregularly applied loads have a far more serious effect on a structure than dead loads or gradually applied loading, but by far the worst damage is caused by regular, rhythmic, repetitive loads which, coupled with the natural resonance of the structure, cause eventual collapse. The tragic loss of the Tacoma Narrows bridge in North-West USA was an example of such a phenomenon and the considerable damage caused to the undercroft and structure of York Minster demonstrates the effect of modern traffic vibrations.

In addition to distortion of the structure caused by the stress and load problems mentioned above, settlement and distortion may have been caused by the action of materials used. These can be broadly divided into defects which occur in the first phase of a building's life – say 25 years from completion – and those which take place at a later date or even have a cumulative effect over the years.

Deterioration or initial shrinkage of materials such as mortar may be responsible for initial deformation of the structure but longer-term effects are caused by weathering, internal or external severe temperature changes, the actions of salts (e.g. sulphate attack or efflorescence), rising damp and changing moisture content in the materials used. Modern buildings use more elastic pointing materials such as mastics and they incorporate movement joints at appropriate intervals if they are well designed. Moisture rising from the site is now well controlled by damp-proof courses and membranes.

Cracking in materials is a regularly occurring phenomenon in all old buildings. Superficial cracking occurs in applied finishes such as rendering and plaster but the more serious faults are caused by the cracking of the structure itself. For example, cracks caused by settlement in the structure or subsidence which may be manifested as raking cracks in a brick wall, initially following the lines of the horizontals and perpendiculars to the mortar joints but eventually cracking the bricks themselves. Brittle stone is often strong in compression but weak in tension, such cracks appearing parallel to the compression force where weakening of the structure has consequently occurred.

Poor quality restoration was apparent in a 'listed' high-rise block seen in the mid-west of the USA which is entirely clad in glazed clay faience blocks and had recently undergone renovation. The renovation seemed to have been applied principally to surface cleaning of the glazed blocks and was largely cosmetic in nature. Many cracks in the glazed surfaces of the blocks with associated crazing of the glaze were evident and it is inevitable that the building envelope has become less than watertight with consequent corrosion of the steel structural frame beneath, exacerbating the cycle of further cracking, more water penetration and further corrosion of the steel frame. It is fortunate indeed that buildings in the USA are generally regarded as expendable with a very limited life.

Compare this with the Inca structures at, for example, Cuzco and Macchu Picchu in Peru. These were built of huge blocks of granite, interlocking, completely devoid of mortar with walls having an out-of-plumb inward batter and it is impossible even to insert a penknife in the joints between the blocks. These buildings were, as a result, so strong that they were virtually earthquake resistant and very permanent in nature.

Although in this country we do not have to design against earthquakes and even minor earth tremors are a rare occurrence, damage to the structure may be caused by excessive water abstraction causing a drying out of the subsoil, on occasion by the collapse of obsolete sewers, damaged water mains causing subterranean cavitation or even – as in Northwich, Cheshire – older houses disappearing into cavities formed by subterranean brine extraction. We are likely to hear increasingly in the future about changes in the water tables causing damage to many buildings, especially in London, as increasing use is made of bore holes as a means of supplying a rising demand for water.

To sum up, the environment in all its aspects must be closely studied to assess the effect of all these factors on the building before commencing the survey itself.

Local techniques of construction must be investigated, as it is most important that the reason why a building has been designed and built in a particular way out of certain materials should be appreciated before reporting on the structural condition. Some rural areas in the UK have a great deal of cob and thatch. The thatched roofs have no eaves gutters, thus dispensing with the rainwater disposal problem; the eaves project further than usual and there is a slow 'run-off' from the semi-absorbent thatch. The 2–3 ft thick cob walls are plastered both sides to prevent ingress and egress of damp, the cob (straw, clay and earth) core forming a moist cohesive 'sandwich'. On no account should a damp-proof course be recommended as the consequent drying out could leave one with a pile of red dust instead of a wall! Rising damp is, however, controlled by building the footing of a cob wall in a semi-absorbent stone such as local sandstone and a very effective method of construction this is, provided the integrity of the internal and external plastering is preserved.

In the Lake District stone walls are laid in thick slate tilted so that the joints slope outwards to prevent damp penetration. Unfortunately in Castle Drogo, Devon – completed around 1930 – the walls are battered and the mortar joints slope inwards resulting in excessive damp penetration. The face of each block should have been battered and the joints made level or sloping outwards. Was this lack of thought or inadequate supervision?

Space does not permit a description of all the many regional variations in the use of materials and forms of construction but the examples given above should be sufficient to show that, before embarking on the survey of an old building, an intensive knowledge of local historical methods of construction and the use of materials is required.

Many regions in this country also have their unique problems covering insect infestation and fungoid infection. The house longhorn beetle is localised in the southern Home Counties and especially the Weybridge area of Surrey. Furniture beetle infestation appears to be more prevalent in the south and south-west than in the north, whereas dry rot seems to spread more rapidly in the north of England than in the south. As older buildings are more likely to suffer structural damage from these causes than are the newer buildings where preventive measures have often been taken, the surveyor must study the potential impact of these factors on the subject property.

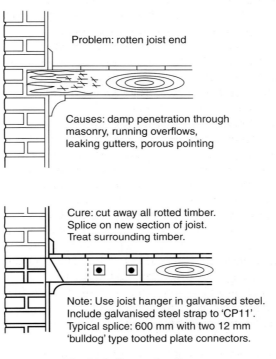

Problem: rotten joist end

Causes: damp penetration through masonry, running overflows, leaking gutters, porous pointing

Cure: cut away all rotted timber. Splice on new section of joist. Treat surrounding timber.

Note: Use joist hanger in galvanised steel. Include galvanised steel strap to 'CP11'. Typical splice: 600 mm with two 12 mm 'bulldog' type toothed plate connectors.

Fig. 21.1 Conserving floors.

When carrying out repairs, traditional materials compatible with the locality should have been used wherever possible but in some cases modern substitutes will have been the only remedy available – for example, stone dust/epoxy resin in stonework repairs. Such substitutes should only have been used as a very last (and extreme) resort where no other means were available. One particular fault often encountered is the repair of gauged brick arches over windows and door openings, often clumsily executed with inferior bricks and sand/cement mortar where soft, sandy 'rubbers' and lime mortar should have been used. As a general rule the use of Portland cement should have been avoided wherever possible in old buildings as it shrinks, is too strong for the surrounding materials and may produce salts which can cause damage.

Planned maintenance and the surveyor

There are two approaches to maintenance: to plan for it or to do the work only when absolutely necessary on an *ad hoc* basis. Effective and regular planned maintenance work should always be a major factor in the conservation process. It is a necessary feature which is the only logical way to ensure the retention of a historic or unique building so that it remains in a usable condition and does not deteriorate.

Floor sagging
and springy

Ceiling cracked
and bulging

Excessive notching-out of old floor joists. Ideally joists
should be drilled. If notching is unavoidable it should
be limited to 1/8th the joist depth. Victorian and earlier
construction often features undersized joists in any event

Cure: slight deflection – insert new joists/additional
joists. Serious deflection – renew whole floor

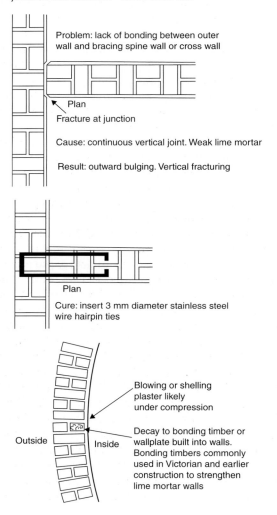

Problem: lack of bonding between outer
wall and bracing spine wall or cross wall

Plan
Fracture at junction

Cause: continuous vertical joint. Weak lime mortar

Result: outward bulging. Vertical fracturing

Plan
Cure: insert 3 mm diameter stainless steel
wire hairpin ties

Blowing or shelling
plaster likely
under compression

Decay to bonding timber or
wallplate built into walls.
Bonding timbers commonly
used in Victorian and earlier
construction to strengthen
lime mortar walls

Outside   Inside

Bulging wall with timber
built into brickwork

Cure: walls seriously out-of-plumb are better
rebuilt reusing bricks when these are sound

Fig. 21.2 Conserving walls.

Planned inspection to a routine can be of use to surveyors in the case of larger, older and unique buildings. It can provide us with invaluable feedback of performance information over a very long period of time, often with materials and forms of construction which are no longer in regular use and not otherwise understood.

Having a planned maintenance system for a building or group of buildings also means that management will sometimes maintain a small, directly employed labour force of specialist tradesmen selected from amongst the older and more experienced craftsmen in the locality, together with a few younger apprentices. Such a system ensures that traditional skills in repair, restoration and renovation are handed down from generation to generation and keeps the continuity in knowledge and techniques alive and is more likely to ensure that the building is kept up to a good standard of repair.

Planned maintenance also requires a reporting system. Neglect of an old building in partial or constant use can be mitigated by briefing staff, caretakers and, in particular, cleaners to ensure that they promptly report telltale signs of want of maintenance. It is often said that the best way of ensuring that no defects occur in one's car is to clean it inside and out regularly by hand, instead of just putting it through the car wash, so that incipient rust and other defects are detected and preventive measures taken in good time. The same principle can apply to a building and very little training is needed to detect damp penetration, broken glass, leaking overflows and gutters, damaged fittings and the like. A chat with the operating staff can often provide the surveyor with invaluable performance information and direct his or her attention to those parts of the property which require a closer inspection.

Another highly valid reason for using all eyes, hands (and sometimes noses) available from the regular staff in a historic or valued building is that such properties are not, as most modern buildings, constructed to last for a limited life but are expected to be of a permanent nature. Any fungoid infection, insect infestation or damp penetration is very likely to have been noticed by operating staff and this information must be acted upon in order to avoid the onset of cumulative damage.

Great care is required in suggesting remedial measures. It is best if possible to ensure that the same methods and materials are recommended as were used in the original construction. The introduction of new 'foreign' materials can sometimes have a serious and even catastrophic effect. Some caution should be exercised when specifying timber treatment having regard, on the one hand, to the need to control woodborers and fungal attack and, on the other hand, the importance of limiting the amount of chemicals used which could have damaging long-term effects on the health of operatives, occupiers and visitors. Vapour barriers must be used with care to prevent undue moisture concentration in sensitive places. The juxtaposition of materials with different characteristics must be carefully considered, for example hard cement mortar should not be used with soft stone or brick, nor untreated softwood fixed to a damp wall.

Vapour barriers are required on the warm side of insulation, but not on the cold side. On the cold side there must be adequate ventilation. Failure to observe these rules when insulating roof voids will result in condensation problems which can give rise to rot and other timber defects. If a roof is recovered in new tiles or slates on battens and sarking felt ventilation must have been included in the specification. Without sarking the void is probably well ventilated but when sarking is added the void is effectively sealed. With insulation added to the ceilings it is sealed and cold. Cut a hole in the ceiling for a service duct or down light – especially over a kitchen or bathroom – and warm damp air is passing through into the cold void, creating a major problem of condensation on the roof timbers. Hence the importance of considering each repair or improvement to a building as part of the total and not in isolation. During the survey inspection it is imperative that any signs of condensation must be thoroughly investigated, the causes traced and remedial measures suggested.

Plumbing, electrical and heating services all have a limited life and require periodic renewal over the lifespan of an ancient building. They therefore need to be accessible and renewable without disturbing the historic structure which means in most cases running cable and pipework on surfaces or within accessible ductwork. Some thought needs to be given to the conflicting requirements of accessibility on the one hand and the desirability of hiding unsightly pipes and cables on the other hand as concealed services do limit the scope and value of an inspection in the case of older buildings.

Although some abandoned buildings and ruins need merely to be kept in both a wind and watertight and structurally safe condition, most buildings to which a conservation policy has applied have received more sophisticated treatment in the form of a quinquennial plan covering major repair, pointing and decoration with regular cleaning out of gutters, gullies, drains, valleys and parapet gutters in particular – say twice yearly. In addition regular routine inspections of security and fire alarm systems, smoke and heat detectors and emergency lighting together with appropriate testing is needed in any important building.

In the case of lofty buildings such as churches, lightning conductors should be installed, inspected, tested and certified – preferably annually. Any certificates should be seen during the course of the survey otherwise an oversight on the part of the surveyor could result in a professional indemnity claim. Remember that it was lightning that caused the disastrous fire to the South Transept of York Minster.

When considering all the foregoing factors the owners' or occupiers' objective should be 'keep, restore or improve every facility in every part of a building, its services and surroundings to an acceptable standard and to sustain the utility and value of the facility'. The 'acceptable standard' for conservation purposes is often dependent upon the amount of finance available as the type of building involved is often administered by a trust, charity, a single owner or is an ecclesiastical building – in short, a 'non-productive' building in that it does not have an income-producing or profit-making function associated with it. This limited maintenance funding in such cases means that very careful pre-planning of resources available must have been carried out including a strong element of preventive maintenance to guard against expensive, catastrophic failures as well as a strict system of priorities. This should have ensured that, for example, structural and roof repairs as well as external painting and gutter maintenance have had priority over internal decorations and work of a purely cosmetic nature.

Maintenance planning for buildings can be assisted using computer-supported databases. Surveyors dealing with maintenance planning and life-cycle costing invariably use a computer database nowadays to store all the relevant information. Any such records of past maintenance work giving details of tasks and dates would be invaluable to the surveyor in assessing the performance of a building.

Many of the cities and towns in developed countries may now be regarded as fully built. The infrastructures are already in place and attention is shifting from construction to conservation with greater emphasis on the economical use of resources and avoidance of waste.

The initial condition survey, including the assembled dimensions and floor plans, is an essential first step in setting up a computer-based maintenance planning system. A well-organised building survey can form the basis from which decisions can be made regarding the future maintenance programme for the structure in question.

Access

Access for survey purposes is often a problem in older buildings and, for example, tower scaffolding is usually necessary to inspect and repair the high-level interiors of cathedrals, churches and halls with consequent extra expense. The provision of walkways and access hatches in roof voids, permanent access and walkways to high-level roofs, hatches to inspect and repair sub-floor areas as well as ducts and accessible trunking for services will all help to reduce future costs and ensure that no part of the structure remains neglected. The more inaccessible the area of a building to be maintained or surveyed (e.g. a church spire or a high steeply pitched roof), the more essential it is that any works should be carried out with a longer-term cycle in mind, even if the cost is greater at the time. One would not wish to have to scaffold the spire of Salisbury Cathedral, for example, more than once in a lifetime at the very most! Perhaps 80–100 years repair-free service would be the figure one would wish to achieve in such a case

Fig. 21.3 London Stock brick set in lime mortar to Flemish bond. Properly formed arches are self-supporting and do not need lintels. Bricklayers who set out the corners are known as 'corner men' whilst those who infill with the main runs of brickwork are known as 'line men'. Corner men are the most experienced and the final appearance of the wall depends very much on their skill.

and a full inspection should be made every time works are carried out making maximum possible use of the access provided.

Further information

Buildings to which a conservation policy applies demand special treatment in the matters which are to be covered by a building survey and the surveyor will need to undertake continuous research and study in order to be able to deal adequately with the problems encountered.

I have not attempted with this volume to write a text book on building construction *per se*; there are many excellent books available on construction techniques and materials, some of which are mentioned in the Bibliography. Much useful information can be obtained from bodies such as the Ecclesiastical Architects and Surveyors Association who, from time to time, publish information sheets on materials and techniques used in relation to cathedrals and churches as well as invaluable guidelines for quinquennial inspections. English Heritage also issue an informative quarterly bulletin and publish books and reports relating to conservation – their current publications relate to subjects as diverse as 'Victorian and Edwardian Stained Glass', 'Directory of Public Sources of Grants for the Repair and Conversion of Historic Buildings' and 'A Farewell to Fleet Street', the last named being a historical and architectural study of Fleet

Fig. 21.4 The arch over first floor window has spread and the lintel has dropped. This section of wall will need to be rebuilt.

Fig. 21.5 The wall around this window leans out and has been restrained by the use of tie bars with the painted tie bar plates used as an added decorative feature. The tie bars will only work if the steel rods are in good order and very firmly secured at the other end.

Street and the London newspaper industry. Other useful literature is available such as 'How Old is Your House?', 'Care for Old Houses' and 'Change of Use', all by Pamela Cunnington, and lastly the entertaining and important 'Blue Guide to Victorian Architecture in Britain' by Julian Orbach.

On a purely practical note, 'Practical Building Conservation', by John and Nicola Ashurst is to be recommended, being published in five volumes as follows:

Vol. 1. *Stone Masonry*
Vol. 2. *Terracotta, Brick and Earth*
Vol. 3. *Mortars, Plasters and Renders*
Vol. 4. *Metals*
Vol. 5. *Wood, Glass and Resins*

These volumes cover the cleaning, repair, use and treatment of the above-named items in detail and are all published by Gower Technical Press of Aldershot.

English Heritage also offers repair grants for selected buildings. Typical examples are a grant of £418 000 for the maintenance, repair and renovation of the Chatham Historic Dockyard and £250 000 to North Tyneside Metropolitan Borough Council for the repair of Tynemouth Railway Station as well as smaller projects such as £66 557 towards roof repairs at the Church of Ascension, Lavender Hill, London. Such works have generally been well planned, closely supervised and information on the scope of work done will be readily available to surveyors generally. The advent of the National Lottery and the possibility of grants from this source may be good news for charities struggling to maintain old buildings.

Although some designers and many developers may say that current concern with conservation represents unhealthy romantic nostalgia for the past – and consequently emasculates technological progress in design and construction – our national culture depends on retaining the best of our living history.

Such is the variation in building techniques in different parts of the country that local knowledge plays an essential part in this type of work and all this presents a fascinating area for study for any keen young surveyor interested in historic buildings.

The conservation bandwagon is now well and truly rolling and surveyors are jumping on board. In 1987 the Royal Institution of Chartered Surveyors (RICS) Conservation Group was formed and 2 years later had attracted 400 members, about half of them specialist building surveyors. One of the aims of the group was to launch a post-graduate diploma in building conservation at the College of Estate Management at Reading. This is now in place so Chartered Surveyors who wish to specialise in the conservation of old buildings have the opportunity to undertake special training.

Not all surveyors will find themselves undertaking a quinquennial survey of a major cathedral or advising on the restoration of a Grade 1 manor house, but on a more mundane level there is still a contribution that we can all make as we undertake our routine work. Valuers and surveyors are front-line professionals who daily come into contact with the nation's building stock. In the process of reporting we can advise and warn about existing defects and also – equally important from a conservation point of view – we can advise about those aspects of the construction on which a little money spent now can save a far greater expenditure later on.

The use of chemicals in conservation

For some time concerns have been expressed about the use of chemicals in buildings as part of the conservation process. Chemicals are used to reduce or prevent rising damp in masonry. Timbers are routinely sprayed with insecticidal and fungicidal fluids to control rot or beetle attack. The operatives using the equipment for chemical injection or spraying and the occupants of buildings being treated may need reassurance.

In 1991 the Health and Safety Executive published a report 'Remedial Timber Treatment in Buildings – A Guide to good practice and the safe use of wood preservatives' with recommendations. Surveyors are advised to obtain a copy. In broad terms the Executive recommends that wholesale treatment of timbers be avoided where possible and that defects such as dry rot and furniture beetle be prevented by good design with less reliance placed on chemical treatment. If timbers are dry and well ventilated most timber defects are avoided and furniture beetle is unlikely to be able to survive in timbers with a moisture content of 12% or less. Widespread rot or beetle attack in a building is an indication that the timbers are damp and that there are design faults.

Global warming and woolly thinking

Having got this far the reader will appreciate some of the complexities of even the simplest buildings and the need to consider the building as a whole and not just parts in isolation; moreover the building has to be considered over its whole life cycle if wasteful short-term expenditure is to be avoided.

Some materials may seem superficially attractive. Unplasticised polyvinyl chloride (PVCu) double-glazed windows are a case in point. Double glazing is heavily promoted now as a means to reduce heat loss and make homes warmer with anticipated savings in fossil fuels. However, well-made wooden windows will last 100 years and the cost of an application of paint or preservative stain every few years will not be great; moreover wood is a naturally renewable material and its production ecologically neutral. Double-hung sashes or casements may well be in keeping with the style and character of an old building.

The production and eventual destruction of PVCu involves industrial processes which are far more harmful; moreover the windows themselves do not last forever and the material eventually yellows and becomes brittle. Double glazing as such takes many years to pay for itself in lower heating bills. The immediate benefits that many property owners report on installing double glazing are often the result of the elimination of draughts rather than the extra sheet of glass.

The double-glazed sealed units also have a limited life. Eventually the seal breaks down resulting in condensation between the panes and the windows then have to be reglazed – the cost of such maintenance, including the manufacture of replacement glass, all forming part of the overall equation.

Fig. 21.6 The grass growing in this gutter results from years of neglect. Sometimes conservation is just a matter of common sense.

22

The surveyor as expert witness

A paradox

If you read the law reports covering those cases in which an issue of valuation was in dispute, and where expert surveyors have given evidence for both sides, you will frequently encounter a curious paradox. In the judgment it will be stated that any reasonably competent surveyor could arrive at a valuation within a band, or range, of value on either side of what may be regarded as the 'correct' valuation. The courts allow for the fact that there will be differences of opinion at any given time between surveyors as to what a particular property is worth. The fact that different surveyors place a different value on a property would not, of itself, suggest that either of them was negligent in arriving at that opinion.

So at various times the courts have decided that there is a band within which a non-negligent valuation will lie. It may be 10%, or 15% or some other figure depending upon the type of property. But it is often quite a narrow band. A valuation falling outside this band may be held to be negligent, or may be found to be negligent if the methodology used to arrive at it was flawed.

However if you look at the expert evidence presented in many of these cases you will find that the experts themselves often propose values which are much wider apart than the band would allow. And these experts are supposedly the top people in their field who have had months, and sometimes even years, to prepare their valuations prior to the court hearing.

So we may find experts differing by 30% in their valuations of a property where it is agreed that a reasonably competent valuer would be expected to value within a 10% band. How can this be?

Three possible explanations come to mind: one or both of the experts are not up to the job, or the concept of a fairly narrow band for a non-negligent valuation opinion is flawed, or one or both of the experts are failing to give the court their unbiased and honestly held opinions.

Surveyors as experts

Suitably qualified surveyors are experts. Their field of expertise is building construction and property valuation so they are fitted to provide forensic and expert evidence in a variety of circumstances including formal court hearings and arbitrations.

There has been a growth in this type of work in recent years in the UK, especially in relation to negligence actions involving building defects and valuations. So surveyors will often be instructed to carry out Building Surveys with a view to preparing reports to be used in litigation.

The actual content of the report or proof of evidence will obviously vary greatly from case to case but the opinions should be presented in the form normally used by most surveyors and advocated in this book, that is, first a description of what was seen; then – if appropriate – an indication of what was not seen; then the conclusions drawn from observation; finally the advice which would normally be given as a consequence of this conclusion.

If the case involves an allegation of negligence against another surveyor then it is invariably necessary to consider in some detail what the circumstances might have been at the time of that surveyor's inspection and to give an opinion as to what a reasonably competent surveyor might have been expected to see, and the conclusions and advice which would have followed, in the circumstances prevailing at the time.

As a general rule it is best to keep the main report or proof short and to add as appendices any plans, sketches, photographs or schedules with reference to these within the main text. Avoid waffle and especially avoid any attempt to indulge in advocacy.

The Civil Procedure Rules

In 1999, following publication of a report by Lord Woolf, new procedures were introduced in England and Wales relating to the conduct of Civil Proceedings. The 'Woolf' reforms were intended to speed up the litigation process and encourage the early settlement of disputes. Protocols have been laid down for the conduct of various classes of proceedings including litigation about building defects and professional negligence. Surveyors active in this area in England and Wales must familiarise themselves with the Civil Procedure Rules and the Protocols which apply.

The Rules and Protocols have considerable effects on the way in which expert evidence is presented and especially stress the importance of dealing promptly with correspondence and requests for meetings from the opposing side. Deliberate or unreasonable delay (a not uncommon tactic in the past) will not work now and can seriously prejudice the client's case. Judges may make an award of costs against a litigant who fails to abide by the Protocols.

The Court may now require the appointment of a single expert, to be agreed by both sides, rather than each side having separate experts. The Court Rules, current at the time of writing, state, *inter alia*:

Rule 35.3

1. It is the duty of the expert to help the court in the matters within his expertise.
2. This duty overrides any obligation to the person from whom he has received instructions or by whom he is paid.

1.1 An expert's report should be addressed to the court and not to the party from whom the expert has received his instructions.

1.2 An expert's report must:
   1. give details of the expert's qualifications,
   2. give details of any literature or other material which the expert has relied on in making the report,
   3. say who carried out any test or experiment which the expert has used for the report and whether or not the test or experiment has been carried out under the expert's supervision,
   4. give the qualifications of the person who carried out any such test or experiment and
   5. where there is a range of opinion on the matters dealt with in the report
      i. summarise the range of opinion and
      ii. give reasons for his own opinion,
   6. contain a summary of the conclusions reached,
   7. contain a statement that the expert understands his duty to the court and has complied with that duty (rule 35.10 (2)) and
   8. contain a statement setting out the substance of all material instructions (whether written or oral). The statement should summarise the facts and instructions given to the expert which are material to the opinions expressed in the report or upon which those opinions are based (rule 35.10 (3)).

1.3 An expert's report must be verified by a statement of truth as well as containing the statements required in paragraph 1.2(7) and (8) above.

1.4 The form of the statement of truth is as follows: 'I believe that the facts I have stated in this report are true and that the opinions I have expressed are correct.'

1.5 Attention is drawn to rule 32.14 which sets out the consequences of verifying a document containing a false statement without an honest belief in its truth.

Rule 32.14 says that proceedings for contempt of court may be brought against a person if he makes, or causes to be made, a false statement in a document verified by a statement of truth without an honest belief in its truth.

So the burden of acting as an expert witness should not be shouldered unless the surveyor is comfortable with the instructions and his or her ability to provide relevant, accurate and appropriate evidence which is truthful and in all respects the honestly held opinions of the expert.

The leading case which sets out the duties and responsibilities of an expert witness is *National Justice Campania Naviera S.A.* v. *Prudential Assurance Co. Ltd* otherwise known as the 'Ikarian Reefer' case and in the Ikerian Reefer case (which predates the Woolf reforms but which is still pertinent) Cresswell J. summarised the duties and responsibilities of expert witnesses in civil cases as follows:

1. Expert evidence presented to the Court should be, and should be seen to be, the independent product of the expert uninfluenced as to formal content by the exigencies of litigation (see also *Whitehouse* v. *Jordan* (1981)).

2. An expert witness should provide independent assistance to the Court by way of objective, unbiased opinion in relation to matters within his experience (see *Polivitti Ltd* v. *Commercial Union Assurance Co. Plc* (1987) and Re. J. (1990). An expert witness in the High Court should never assume the role of advocate.

3. An expert witness should state the facts or assumptions upon which his opinions are based. He should not omit to consider material facts which could detract from his concluded opinion (see Re. J. (1990)).

4. An expert witness should make it clear when a particular question or issue falls outside his expertise.

5. If an expert's opinion is not properly researched because he considers that insufficient data is available, then this must be stated with an indication that his opinion is not more than a provisional one. In cases where an expert witness who has prepared a report cannot assert that the report contains the truth, the whole truth and nothing but the truth without some qualification should be stated within the report (see *Derby and Co. Ltd and others* v. *Wheldon and others* (1990)).

6. If after exchange of reports an expert witness changes his view on a material matter having read the other side's expert report or for any other reason, such changes of view should be communicated (through legal representatives) to the other side without delay and, when appropriate, to the court;

7. When expert evidence refers to photographs, plans, calculations, analysis, measurements, survey reports or other documents, these must be provided to the opposite party at the same time as the exchange of reports (see 15.5 of the Guide to Commercial Court Practice).

The RICS Practice Statement and Guidance Notes

In 1997 the Royal Institution of Chartered Surveyors (RICS) published 'Surveyors Acting as Expert Witnesses – Practice Statement and Guidance Notes' which came into force on 1 March 1997 and which has been periodically updated and reviewed since to accommodate the Woolf reforms and other matters. The document is in two parts both of which have applied from 1 March 1997 to any Chartered Surveyor who agrees or is required to provide expert evidence, whether oral or in writing, which may be relied upon by any Judicial or quasi-Judicial body in the UK.

In respect of the surveyor's written evidence the Practice Statement requires the surveyor to include a declaration, which may be qualified:

(i) of belief in the accuracy and truth of the matters put forward;
(ii) that the report includes all facts which the surveyor regards as being relevant to the opinion which he has expressed and that he has drawn to the attention of the Judicial Body any matter which would affect the validity of that opinion and
(iii) that the report complies with the requirements of the RICS as set down in 'Surveyors Acting as Expert Witnesses: Practice Statement'.

Some practical considerations

In relation to Building Surveys the initial instructions are likely to come from a solicitor acting for a client contemplating, or already engaged in, litigation. These instructions will ask for an inspection of the property and a preliminary report dealing with the matters in dispute which may relate to one or more specific defects in the building and/or a question of value. It may be necessary to carry out a survey of the whole property, or it may be sufficient to inspect particular parts.

At this stage the surveyor should confirm precisely what the instructions are and their purpose. He or she should agree the level of fee and manner – and timing – of payment. It is also essential in my view that the surveyor should know the identity of all the parties involved in the dispute in case there may be a potential conflict of interest. Also if the litigation is to involve an allegation of negligence relating to a surveyor's report, a valuation or other advice given then all such reports or other documents must be provided at the outset.

I have encountered circumstances where a solicitor may ask for a report on a property in contemplation of litigation without providing full background details. It can be very dangerous for an expert surveyor to proceed on this basis. Later it may be found that a connection arises with one of the parties leading to a conflict of interest, or a preliminary report may be prepared leading to conclusions which are found later to need changing in the light of the contents of documents which the surveyor had not seen at the time.

In negligence cases a fellow professional's reputation is on the line. Significant claims for damages and costs may be involved. So whether you are acting for the claimant, defendant or as a single independent expert appointed by the Court some time should always be spent ensuring that you are fully aware of all relevant aspects of the case before expressing your own opinion.

Inspections

The building being inspected may be occupied by one of the parties to the litigation, but often it is not. It is not unusual to find that the litigation is about a report prepared or work undertaken some years ago, and the property may have changed hands. If the litigation involves a bank or other mortgage lender then the property may have been taken into possession and sold.

In practice it is often necessary to offer the occupants money to permit an inspection of their property and I have found the going rate at the time of writing in the London area to be £50. Occupants of properties sold by mortgagees in possession are frequently reluctant to permit inspection since they may have been in receipt of unwanted mail or visits from people chasing money from the previous owners. If a request comes from a surveyor wishing to inspect the house in connection with some litigation they may prefer not to become involved. The offer of some cash in such circumstances usually does the trick. Certainly if someone offered the author £50 to look at his house he would be very happy to take the money.

Preliminary reports

Preliminary reports and other correspondence between the surveyor and the instructing solicitor will normally be privileged – that is to say the opposing side in the litigation will have no right to see them under the normal rules of disclosure under which documents are exchanged between claimant and defendant prior to the court hearing.

As a general rule therefore any such reports should be headed 'DRAFT' and 'PREPARED IN CONTEMPLATION OF LITIGATION' to ensure that there is no confusion about their privileged status.

Only when a report is ready, in its final form, for exchange with the other side or submission to the Court (in the case of a single appointed expert) should these headings be removed. It is also important to ensure that a report which is exchanged does not refer within its text to other privileged documents since if the parties see such references they may be entitled then to the disclosure of the documents referred to.

Conferences with counsel

Under the previous, more adversarial, procedures a surveyor as expert witness would be required to attend meetings with the solicitor and barrister – usually at the barrister's chambers – in order to consider in detail the contents of the expert report. Such conferences with counsel would include careful questioning of the surveyor in order to gauge how he or she might perform in the witness box.

Following the Woolf reforms such meetings are still held but the surveyor now has an extra degree of independence and whilst suggestions may be made as to how the report could read better or be better arranged the lawyers cannot attempt to influence the report or allow what Cresswell J. called 'the exigencies of litigation' to result in a report which is partisan or biased.

Meetings of experts

During the course of civil proceedings, if a single expert is not appointed, then it is quite common for the judge to direct that there be a meeting of the expert witnesses for the purpose of agreeing facts. This saves time in court. In connection with buildings it would be common practice for the surveyors acting as experts on each side to meet and agree as much basic information as possible and also identify those areas where there is no agreement.

Such meetings are normally held on a 'without prejudice' basis – that is to say the content of the discussions is not disclosed to the Court until such time as minutes of the meeting are agreed which can be formally placed before the judge.

A surveyor attending such a meeting must be careful to confine the discussion to the instructions given, must avoid any temptation to engage in an exercise in advocacy and should not treat the meeting as an opportunity for negotiation. The object is to agree as many facts as possible and agree to disagree on those areas in dispute.

Draft minutes should be prepared, exchanged and agreed between the surveyors.

Giving evidence in Court

On arrival at the Court the surveyor should make him/herself known to the instructing solicitor and counsel and also to the Clerk of the Court indicating if he or she has any particular requirements in connection with swearing in. Christians will swear the familiar oath on the Bible but alternative arrangements exist for those of other faiths to swear appropriately and for those of no faith to affirm if they wish. The Clerk to the Court will deal with this.

An expert appointed by the Court should sit in a neutral position but experts appointed by one side or the other will normally sit behind the instructing solicitor and counsel, if possible within speaking or note-passing distance. I find it preferable to be in Court for the whole proceedings and not just for the period during which I give evidence so as to be fully familiar with the other evidence given together with any cross-examination or re-examination which has followed. It is quite common for counsel to consult the expert during the trial on technical points.

When called to give evidence in the High Court the surveyor will be sworn in and will initially be asked some questions by counsel for his or her side. The report will already be entered in evidence and will have been seen by the judge so the expert is not required to read the report (as is sometimes the case in planning enquiries and the like). The initial questions are called the 'examination in chief' and are generally brief, intended to confirm the identity of the expert and verify that the copy of the report in the document bundles is the expert's report.

There follows cross-examination for the other side and counsel for the other side may attempt to undermine the expert's credibility and the content of the expert's report. When the expert is appointed jointly as a single expert it is possible that counsel for both sides might wish to probe the conclusions and test the opinions. If the expert is confident and the conclusions in the report are the honestly held opinions of the expert, arrived at following careful research and reflective thought then there should be nothing to fear from this part of the process.

After cross-examination opposing counsel may then ask some further questions by what is called re-examination. Finally the judge may ask questions of the expert. In my experience the judge's questions are often the most relevant and one can frequently gauge from the judge's questioning the way in which the judge's mind is working.

I should add at this stage that it is obviously important to speak clearly when giving evidence and the replies to all questions in court should be given facing towards the judge rather than the barrister asking the question, speaking slowly if the judge is making notes.

Arbitration and other quasi-judicial hearings

The forgoing is a brief summary of the procedures which may apply when a surveyor is giving expert evidence in the High Court in England and Wales. There are many other types of hearing where expert evidence on building construction or property valuation matters may be required.

Generally arbitration hearings and the like are much more informal. The author gave evidence some years ago at an arbitration hearing in respect of a claim under the National House Building Council scheme in relation to defects in a new house. The hearing – by a local architect as arbitrator – was held in the dining room of the house with the property owner, the builder and experts for both sides sitting around the dining room table with the arbitrator. We were fed copious quantities of tea, biscuits and home-baked cakes by the owner's wife.

At some quasi-judicial hearings the surveyor can act as both expert witness and advocate for the client wearing – as it were – two hats. In its Practice Statement and Guidance Notes the RICS has something to say about this.

The RICS says that when the surveyor assumes the roles of both expert and advocate – in proceedings where this is permissible – the roles must be clearly distinguished and declared to the Judicial Body and the opposing party. Circumstances where such a dual role may be permitted include Valuation Tribunals and Lands Tribunal, Planning Enquiries, Arbitrations, hearings before Independent Experts, Rent Assessment Committees and Leasehold Valuation Tribunals.

In *Multimedia Productions Ltd* v. *Secretary of State for the Environment and another* (1988) EGSC, a case concerning a planning appeal, the judge warned that 'although there is no rule against a person combining the roles of Expert and Advocate before a Public Local Enquiry it was an undesirable practice. An Expert

Witness has to give a true and unbiased opinion; the Advocate has to do the best for his client. The Expert who has also played the role of Advocate should not be surprised if his evidence was later treated by a Court with some caution.'

A particular problem is that when an Expert Witness is giving evidence and there is an adjournment, perhaps overnight or for lunch, the witness cannot discuss the case with the client during the adjournment as he or she is still under oath. During an adjournment a witness can talk about anything other than the case in hand. Any discussion of how the case is going or any other matter connected with the case is forbidden. So the Expert cannot resume the role of Advocate during adjournments and discuss the case.

And finally . . .

Perhaps there may be less – rather than more – expert witness work for surveyors in the future if all of us in the profession can work towards higher standards in surveying and reporting. So we may then have fewer disgruntled or litigious clients and less need to appear in court. If this volume will help towards that end then I am grateful for the opportunity to have written it and to you – the reader – for having read it to the end.

Glossary

Aggregate   Material mixed with portland cement to form concrete. Fine aggregate is sand, course aggregate is gravel.

Air bricks   Perforated earthenware airvents to provide sub-floor or roof void ventilation.

Anaglypta   Thick embossed lining paper for walls and ceilings.

Anobium punctatum   The common furniture beetle, generally the most usual form of woodworm in the UK.

Architrave   Joinery moulding around door or other opening.

Arris rail   Triangular in section timber used to support close boarded fencing.

Artex   Decorative textured coating for walls and ceilings.

Asbestos   Fibrous mineral with fire-resistant qualities. Airborne fibres are a known health hazard.

Ashlar   Close fitting square cut building stones, often of thin section used as a facing to other materials.

Asphalt   Mastic asphalt applied hot in semi-liquid form as waterproof coating to flat roofs, basement tankings and exterior stairways.

Back addition   Projecting rear wing of house, termed an outrigger in some areas.

Back land   Site with no road frontage surrounded by other development or land in other ownership.

Balanced flue   For a room sealed boiler which takes oxygen from the outside air and not from within the room in which it is located.

Ballast   Sand and gravel mixture used for making concrete, etc.

Balustrades   Staircase and landing handrails and spindels.

Barge board   Sloping timber fascia to roof verge or gable end.

Bark borer   Woodworm found only in bark and sapwood, generally harmless.

Battens   Narrow timber strips supporting tiles on slates on rafters.

Benching   Cement work around channel in base of manhole.

Binder   Roof timber fixed over joists in other direction as bracing.

Birdsmouth   Triangular cut out of roof strut to tightly wedge purlin.

Bitumen   Tar like material used in sealants, mineral felts and damp proof course.

Black-ash mortar   Industrial ash used instead of sand with cement and lime.

Blown   Defective render of plaster lifting from base, hollow and loose.

Bonding   Various patterns for laying bricks to maximise wall strength.

Borrowed light   Window in internal wall between rooms, often over door.

Breather membrane   Timber-frame construction wall membrane allows moisture to escape.

Breeze block   Building blocks made of cinders and cement used for internal partitions and inner skin of cavity walls.

Bressumer   Beam spanning opening supporting construction above.

Building Regulations   Statutory control over new building works and alterations.

Building Survey   Formerly Structural Survey.

Calcium chloride   Additive mixed in concrete may result in loss of strength.

Calcium silicate bricks   Subject to thermal expansion and contraction resulting in cracking.

Calorifier   Heating coil of pipework within copper hot water cylinder.

Capillary action   Upward movement of moisture in walls and floors.

Carbonation   Loss of strength to concretes associated with chemical changes and rusting to steel reinforcement.

Casement   Opening hinged part of window.

Cast-*in-situ*   Concrete or other material cost on site within timber or other formwork.

Caulking   Sealing to edges around baths and showers.

Cavity wall   Two skins of masonry with air gap in between.

Cess pit   Sealed tank holding sewage requiring periodic emptying.

Cheek   Side face to roof dormer or bay.

Cob   Thick rendered walls built of earth or clay often mixed with straw.

Code of Measuring Practice   Royal Institute of Chartered Surveyors (RICS) recommended rules for calculating floor areas, etc.

Codes of Practice   Non-statutory recommendations for the use of building materials and techniques.

Collar   High level roof timber running between opposing rafters to prevent spread.

Combination boiler   'Combi' boiler provides heating and hot water directly from mains with no need for storage tanks.

Common furniture beetle   Woodworm commonly encountered in older UK buildings.

Comparables   Other properties sold or values to which reference is made when valuations are prepared.

Condensation   Water condenses on surface when it is colder than the dew point of the surrounding air.

Coniophora puteana   A common form of wet rot fungus.

Consumer unit   Fuse or circuit breaker box controlling electricity supply.

Continuity of cover   Insurance cover against subsidence or other risks carried on from one property owner to the next.

Conventional flue   Boiler takes oxygen from air in room in which it is located with combustion gasses discharged via flue or chimney.

Conversion (1)   Property now used differently, e.g. flat within former house.

Conversion (2)   Chemical change in concrete or other material leading to loss of strength.

Coping   Brick, stone or tile finish to top of parapet.

Corbel   Projection on face of wall to support end of purlin, beam, etc.

Core sample   Drilled out section of concrete or other material taken for analysis.

Cornice   Decorative plaster or rendered moulding at junction of wall and ceiling or below external parapet.

Coving   Curved or shaped plaster finish between wall and ceiling.

Cowl   Shaped chimney pot or terminal on flue to improve efficiency.

Creep   Spreading and folding of lead or asphalt on roofs and steps especially due to heat from sun.

Cross wall   Wall running from side to side.

Cruck   Irregular sections of tree trunk used for rafters and other rough carpentry.

Curtain wall   Lightweight thin outer panel wall.

Curtilage   Enclosed garden area belonging to dwelling.

Dado   Lower part of internal wall to approximately 1 m high usually finished with Dado Rail in form of timber moulding.

Damp proof course   Brick, lead, felt or metal layer at base of wall to resist rising damp.

Damp proof membrane   Layer of polythene, bitumen or other material in solid floor to prevent rising damp.

Death watch beetle   Large woodworm found in damp oak and other hardwoods.

Dentil   Tile fillet to seal joint at base of stack or parapet.

Detailing   Flashings, upstands, soakers and other roof joint weather sealing.

Dishing   Sagging to centre of floor or roof slope.

Dormer   Window projecting out from roof slope.

Double glazing   Sealed units have two panes of glass factory sealed; secondary double glazing has additional window fixed to main window, usually inside.

Dry rot   Most damaging form of fungal attack to damp and also adjacent dry wood with propensity to spread within roofs, floors and panelling.

Easement   Right over adjoining property for drainage or other services, etc.

Eaves   Overhanging base to sloping roof.

Efflorescence   Salt deposits on masonry or roof tiles where dampness evaporates.

Electro-osmosis   Proprietory system for preventing rising damp by electrically earthing wall.

Endoscope   Borescope for inspecting inside wall cavities, etc.

Eyebrow window   Set into roof slope under curving rows of tiles.

Facade   Front elevation of building.

Fascia board   Flat board at eaves used to fix gutter.

Fibreboard   Soft porous building board used for insulation and lining.

Fillet   Sealing of joint at corners using mortar.

Finlock gutters   Proprietory name for interlocking concrete gutter system.

First fixing   Installation of services and fittings prior to plastering.

Flank   Side elevation.

Flashing   Metal covering to joint of roof and stack or parapet, etc.

Flaunching   Bedding mortar to base of chimney pots.

Flight   Straight run of staircase.

Flying freehold   In England and Wales the ownership of airspace over another freehold.

Flying shore   Temporary support in gap between buildings, generally during redevelopment.

Foundations   The below ground construction supporting the walls.

Frog   Recess in brick to hold mortar.

Gable   Triangular upper part of wall from eaves to ridge.

Gallopers   Temporary timber struts under converging chimney brickwork in roof space.

Galvanic corrosion   Rusting of galvanised steel in presence of lead and copper.

Gang   Number of sockets or switches.

Gang nailed trusses   Prefabricated roof timbers fixed with metal plates.

Gauged brickwork   Arches and ornamentation in shaped or tapered bricks.

Ginnel   Alleyway or passage between buildings, mainly north of England.

Going   Staircase distance between risers.

Gravity circulation   Wide bore heating circulation without pump.

Grout   Filling of joints in paving and tiling.

Gullies   Exterior drains into which water discharges.

Gypsum   Modern plaster material used in plasterboard and for plaster skim.

HAC   High-alumina cement additive used in concretes.

Hardcore   Broken brick, stone, concrete.

Haunching   Cement work used to support drainwork and manholes below ground.

Header   Brick laid with end showing.

Heave   Lifting of foundations due to clay swell or other expansion of support below.

Herringbone strutting   Timbers laid in X-pattern between joists.

Hip   Junction of two roof slopes.

Hip hook   Metal bracket holding bottom hip tile in place.

Hoggin   Hardcore.

Hogging   Hump in floor.

Homebuyer Report   Pro forma type survey report.

Honeycombe wall   Bricks laid with gaps to allow ventilation.

Hopperhead   Collects water at the top of a down pipe.

House longhorn beetle   Destructive type of woodworm, mainly confined to southern counties in UK.

Infill   Hardcore laid under solid floor.

Interceptor   Trap in drain.

Invert   Bottom of manhole or drain. Invert level is distance below ground.

Interstitial condensation   Trapped dampness in airspace in wall or roof.

Jamb   Vertical side face to window or door opening.

Joist   Timber running parallel to ground supporting floor or ceiling.

King post roof truss   Traditional design with central post (Queen post has two side posts).

Lamination   Lifting of surface to tiles, bricks, plaster, etc.

Lath and plaster   Thin timber strips with wet plaster coatings.

Lintel (Lintol)   Beam over window or door opening.

Made ground   Difficult sites to build on containing rubbish, hardcore, fill, etc.

Maisonette   A flat on more than one level often with own external entrance door.

Mansard roof   Has steep slope at bottom often with windows and shallow slope above.

Meggar test   Electrical insulation check.

Messuage   Dwelling house and curtilage.

Microbore heating   Narrow flexible pipework.

Mineral felt   Flat roof covering, usually bitumen based.

Moisture meter   Measures electrical conductivity in walls or wood and hence dampness.

Monitoring   Observing crack damage and other movement over time.

Mono pitch roof   Has only one slope from high wall to low wall.

Motorised valve   Electrical controls on hot water and heating pipework.

Movement joint   Gap in masonry wall with flexible sealant to allow movement.

Mullion   Vertical division of panes in window.

Mundic blockwork   Masonry construction using pre-cast blocks made from mining waste.

Nail sickness   Slates slip due to rusting of nails holding them.

Needles   Temporary cross beams used to support construction above.

Newel   End post to balustrade or staircase handrail.

N.H.B.C.   National House Building Council issues certificates for new homes.

No-fines concrete   Cast-*in-situ* concrete walls and floors with no sand aggregate.

Noggins   Short horizontal timbers between vertical studs.

Nosing   Edge of step or sill.

Offshoot   Rear wing to house, north of England.

One pipe heating   Central heating water runs in and out of each radiator in turn.

Oriel   Underside of projecting window bay, shaped or rounded.

Outrigger   Rear wing to house, north of England.

Oversailing   Brick or tile courses projecting out from face of wall or chimney.

PRC   Pre-cast reinforced concrete.

Pad and beam   Foundation comprising concrete blocks set into ground linked by beamwork.

Pantiles   Undulating shaped interlocking tiles.

Parapet   Top of wall at roof or balcony level.

Parging   Cement lining around inside of chimney flue.

Party wall   On boundary between properties in separate ownership.

Parquet   Tongued and grooved, secret nailed, hardwood flooring.

Pebbledash   Cement render to walls covered with pebble.

Pegtiles   Handmade clay tiles held onto battens by wooden pegs.

PI Insurance   Professional indemnity cover for negligence.

Pier   Vertical column of masonry to add strength to wall.

Piles   Concrete columns driven or cast in subsoil as foundations.

Pine end   Flank wall. South Wales.

Pitched roof   Sloping tile or slate roof as opposed to flat roof.

Plant   Builders' tools and heavy equipment.

Plasterboard   Gypsum plaster sandwiched between two sheets of cardboard.

Plasticiser   Added to mortar to improve workability.

Plasticity tests   To determine if subsoils are shrinkable with varying water content.

Plate   Horizontal timber on wall to spread load of joist and rafter ends.

Plinth   Widening at base of wall, typically cement rendered.

Plumb   Vertical. Hence plumb line to test for verticality.

Pointing   External face of mortar between bricks.

Ponding   Water lying on flat roofs.

Powder post beetle   Fairly rare very destructive woodborer.

Pressure grouting   Repairs to drains or sunken floor slabs using pumped cement slurry.

Purlin   Roof timber running sideways under rafters.

Purpose-built   Constructed for current use, i.e. not a conversion.

Queen post   Traditional roof truss system with two off centre upright posts.

Quoin   External corner of wall.

Racking   Sideways shift to timber-frame wall or roof.

Radon   Radioactive gas emitted from some sites especially granite rocks.

Raft foundation   Concrete slab supports walls.

Rafter   Sloping roof timber supporting roof slopes.

Raking shore   Diagonal support to wall from site, usually temporary.

Red Book   RICS manual containing codes of valuation practice.

Rendering   Mortars applied to face of wall.

Retaining wall   Holds back land behind and may thus support structures behind also.

Reveal   Inner face of corner to door and window openings.

Ridge   Top of pitched roof.

Rise   Vertical spacing between stair treads.

Riser   Vertical front of step.

Rising damp   Capillary action causes moisture to rise up in wall or floor.

Rolled steel joist   Mild steel section commonly used over openings in load bearing walls.

Rotation   Tilting movement of foundation or wall.

Roughcast   Unevenly finished external render.

Sacrificial anode   Aluminium or other metal in galvanished water tank to reduce corrosion.

Sarking   Underfelting below tiles or slates.

Sash   Sliding inner frame to window.

Scantling   Dimensions of roof timbers.

Schedule of condition   Prepared as evidence at start of lease.

Schedule of dilapidations   Landlord's claim for breach of covenant to repair.

Scott schedule   List of dilapidations and costings as part of claim for court purposes.

Screed   Cement finish to concrete floor.

Scrim   Hessian type material used to seal joints in plasterwork.

Second fixing   Completion of fittings and services after plastering.

Self-priming cylinder   Hot water tank without header tank.

Septic tank   Private drainage system uses bacterial action to process sewage.

Serpula lacrymans   The true dry rot fungus.

Sets   Small stones used for paving.

Settlement   Downward movement of structure on site.

Shakes   Splits and cracks in timbers following grain.

Shear   Vertical crack due to part of wall moving down.

Shelling   Breaking away of surface to plasterwork or other finish.

Shingles   Thin timber tiles used for roofs and wall cladding.

Shiplap   Overlapping boarding as cladding to external face of wall.

Shore   Steel or timber prop used as temporary support.

Sleeper wall   Under timber ground floor used to support joists.

Slenderness ratio   Calculation of height, width and strength of wall.

Sleugh   Land drain with pipes butted together.

Snagging   Minor building works to be finished off after practical completion.

Snap headers   Half bricks in outer skin of cavity wall replicates solid wall bonding.

Soakaways   Land drains and sumps allowing water to drain into soil.

Soakers   Metal dressing under tiles and flashings.

Soffit   Undersides of eaves behind gutter.

Soil stack   Above ground pipework taking waste water.

Soldier arch   Bricks laid on end as lintel to opening.

Spalling   Flaking away of surface to bricks and tiles.

Spandrel   Space below staircase or over an archway.

Spardash   Cement render finished with small white stones.

Spindels   Vertical uprights to balustrades and staircase handrails.

Spine wall   Internal wall running front to back.

Spouting   Rain water gutters and downpipes, north of England.

Sprocket   Angled timber at end of rafter lifts bottom tiles.

Stitch bonding   Repair to brickwork, cutting in new bricks.

Stramit   Insulation boarding made from compressed straw.

Stretcher   Brick laid sideways.

Strings   Timbers supporting ends of stair treads and risers.

Structural survey   Building survey.

Strutting   Angled timbers supporting purlins and rafters.

Stucco   Smooth cement rendering as external finish.

Stud partition   Timber frame clad in plaster.

Subframe   Outer part of window frame fixed to wall.

Subsidence   Downward shift of building due to movement in ground beneath.

Subsoil   Material below topsoil which supports foundations.

Sulphates   Chemicals which can cause deteriorating in concretes.

Swallow hole   Collapse of subsoil often chalk.

Tanking   Waterproof lining to cellar or basement.

Tell-tale   Gauge fixed over crack to monitor movement in wall.

Thermoplastic tiles   Rigid plastic tiles used in 1950s.

Threshold   Sill to exterior door opening.

Tie bar   Metal restraint inserted in buildings with end plates on wall.

Timber-frame houses   Load bearing inner skin timber framed.

Tingles   Metal clips holding loose slates.

Torching   Mortar applied under slates or tiles to improve weather tightness.

Transom   Horizontal division in window.

Trap   Bend in waste pipe prevents air from drain passing upward.

Tread   Step in staircase.

Truss   Timber framework as part of roof construction.

Underpinning   Insertion of new foundation beneath existing foundation.

Upstand   Vertical face of flashing or soaker.

Valley   Lower intersection between two roof slopes.

Vapour check   Polythene or similar material prevents warm damp air entering void.

Vendor survey   Seller commissions surveyor's report when offering house for sale.

Verge   Sloping edge to gable end wall.

Vertical damp proof course   Used at change of level, around openings to cavity and in basements.

Wainscot   Wooden panelling below dado height around room.

Wall plate   Timber on top of wall supports rafters and joists.

Wall tie   Metal fixing in cavity wall connecting two skins.

Wattle and daub   Early type of plaster finish for internal walls.

Weatherboard   Overlapping boarding used as external wall cladding.

Weep holes   Allow drainage from wall cavity or from behind retaining wall.

Wet rot   Fungal attack to damp wood especially exterior timbers.

Woodworm   Furniture beetle and other wood-boring insects.

Xestobium rufovillosum   Death Watch Beetle.

Bibliography

'Handling of Arrears and Possessions: CML Statement of Practice', Council of Mortgage Lenders.

'Shelling of Plaster Finishing Coats', Building Research Establishment (BRE) Information Sheet, TIL 14, 1971.

'Settlement of a Brick Dwellinghouse on Heavy Clay, 1951–1973', W. H. Ward, BRE, CP 37/74, February 1974.

'High-Alumina Cement Concrete in Buildings', BRE, CP 34/75, April 1975.

'A Simple Pull-Out Test to Assess the Strength of In Situ Concrete', BRE, CP 25/77, June 1977.

'Field Measurement of the Sound Insulation of Plastered Solid Brick Walls', E. C. Sewell and R. S. Alphey, BRE, CP 37/78, March 1978.

'Foundations for Low-Rise Buildings', BRE, CP 61/78, September 1978.

'The Performance of Cavity-Wall Ties', BRE Information Paper, IP 4/81, April 1981.

'House Longhorn Beetle Survey', R. Geraldine Lea, BRE Information Paper, IP 12/82, July 1982.

'Inspection and Maintenance of Flat and Low-Pitched Timber Roofs', W. T. Hide, BRE Information Paper, IP 15/82, August 1982.

'Considerations in the Design of Timber Flat Roofs', I. S. McIntyre and D. P. Birch, BRE Information Paper, IP 19/82, September 1982.

'Assessment of Damage in Low-Rise Buildings', BRE Digest, 251, July 1981 (HMSO).

'The Durability of Steel and Concrete', Parts 1, 2 and 3. BRE Digest, 263, 264 and 265, July, August and September 1982 (HMSO).

'Specifications for Roofing Felt', British Standards Institution (BSI), BS 747, 1977.

'Metal Ties for Cavity-Wall Construction', BSI, BS 1243, 1978.

'Guide to the Choice, Use and Application of Wood Preservatives', BSI, BS 1282, 1959, Revised 1975.

'Preservative Treatment for Constructional Timber', BSI, BS 5268, Part 5, 1977 (supersedes CP 98).

'Code of Practice for Trees in Relation to Construction', BSI, BS 5837, 1980.

'Foundations and Sub-structures for Non-industrial Buildings of Not More Than Four Storeys', BSI, CP 101, 1972.

'Structural Recommendations for Load-Bearing Walls', BSI CP 111, 1970.

'The Structural Use of Timber', BSI, CP 112, Part 2, 1971.

'Roof Coverings: Mastic Asphalt', BSI, CP 144, Part 4, 1970.

'Small Sewage Treatment Works', BSI, CP 302, 1972, 1st revision. Revised as 'Code of Practice for Small Sewage Treatment Works and Cess-Pools', BS 6297, 1983.

'Foundations', BSI, CP 2004, 1972.

'Building Regulations, and Building Regulations (Amendments)', Her Majesty's Stationery Office (HMSO).

'Common Defects in Buildings', H. J. Eldridge, Property Services Agency (PSA), Department of the Environment (DoE), HMSO, 1976.

Small Claims in the County Court, HMSO.

Quality in Traditional Housing (Series), R. B. Bonshor and H. W. Harrison, HMSO, 1982: Vol 1: *Investigation into Faults and Their Avoidance*; Vol 2: *Aid to Design*; Vol 3: *Aid to Site Inspection.*

Registered House-Builders' Handbook, and Practice Notes, NHBC.

'Timber-Framed Buildings', NHBC Practice Note 5 (1982), Rewritten and Reissued November 1982.

'Suspended-Floor Construction for Dwellings on Deep-Fill Sites', NHBC Practice Note 6, August 1973, Revised April 1974.

'Structural Surveys of Residential Property', Royal Institution of Chartered Surveyors (RICS), Practice Note (Surveyors' Technical Services (STS)).

Buying a House? RICS (STS).

Guide to House Rebuilding Costs for Insurance Valuation (annual), RICS Building Cost Information Service (BCIS) with British Insurance Association.

'Code of Measuring Practice', Jointly RICS/ISVA (Incorporated Society of Valuers and Auctioneers), 5th edn, 2001.

Manual of Timber-Frame Housing (A Simplified Method), (National Building Agency) NBA/TRDA (Timber Research and Development Association), The Construction Press, 1980.

'Structural Recommendations', *Timber-Frame Housing Design Guide*, Section 9, TRDA, 1979.

Trees in Britain, Europe and North America, Roger Phillips, Pan Books Ltd, 1978.

'Building Defects', *The Chartered Surveyor*, Supplement, April 1981.

'Regulations for Domestic Equipment', Institution of Electrical Engineers (IEE).

Mitchell's Building Construction (Series), B. T. Batsford Ltd, Vol 1: *Structure and Fabric*, Part 1, 2nd edn, 1979; Vol 2: *Structure and Fabric*, Part 2, 2nd edn, 1977; Vol 3: *Materials*, 2nd edn, 1978; Vol 4: *Components*, 2nd edn, 1979.

Building Techniques (Series – Science Paperbacks), Chapman & Hall, Vol 1: *Structure*, 2nd edn, 1979: Vol 2: *Services*, 3rd edn, 1980.

Principles of Element Design, Peter Rich, George Godwin Ltd, 1977.

Design Failures in Buildings (Series), First Series 1972, Second Series 1974, George Godwin Ltd.

The Law and Practice of Arbitrations, John Parris, George Godwin Ltd, 1974.

Builders' Detail Sheets (Series 2), Northwood Publications, 1977.

Architects' Standard Catalogues (Annual) Standard Catalogue Information Services Ltd.

Spon's Architects' and Builders' Price Book (Annual), E. & F. N. Spon.

Remedial Timber Treatment in Buildings – A Guide to Good Practice and the Safe Use of Wood Preservatives', HMSO Publications, 1991.

Related books from Butterworth-Heinemann, Architectural Press and Laxton's

Altering Houses and small-scale Residential Development, A Bridger and C. Bridger, 1998.

Architect's Legal Handbook: The Law for Architects, 7th edn, eds. A. Speaight and G. Stone, 2000.

Basic Surveying, 4th edn, R. E. Paul and W. S. Whyte, 1997.

Building Adaption, J. Doughlas, 2001.

Building Construction Handbook, 4th edn, R. Chudley and R. Greeno, 2001.

Building Design Management, C. Gray and W. Hughes, 2001.

Building in Value: Pre-design Issues, eds. Rick Best and Gerard de Valence, 1999.

Computing in Construction: Pioneers and the Future, Rob Howard, 1998.

Conservation of Brick, J. Warren, 1998.

Conservation of Building and Decorative Stone, J. Ashurst and F. G. Dimes, 1998.

Conservation of Historic Buildings, 3rd edn, Sir Bernard Fielden, 2003.

Contract Practice for Quantity Surveyors, 3rd edn, J. Ramus and S. Birchall, 1996.

Dampness in Buildings: The Professionals and Home Owners Guide, 2nd edn, E. G. Gobert and T. Oxley, 1994.

Dissertation Research and Writing for Construction Students, S. Naoum, 1998.

Ecology of Building Materials, B. Berge, 2000.

Economic Analysis for Property and Business, Marcus Warren, 2000.

Estimating and Tendering for Construction Work, 2nd edn, Martin Brook, 1998.

Estimating for Builders and Quantity Surveyor, Eric Flemming, R. D. Buchan and Fiona Grant, 2003.

Historic Floors, J. Fawcett, 2001.

Laxton's Building Price Book. Major and Small Works, ed. D. Adler, 1999.

Metric Handbook: Planning and Design Data, 2nd edn, ed. D. Adler, 1999.

Remedial Treatment of Buildings, 2nd edn, B. Richardson, 1995.

Sir Banister Fletcher's A History of Architecture, 20th edn, ed. D. Cruickshank, 1996.

Sustainable Architecture, Brian Edwards, 1998.

Victorian Houses and their Details, H. Long, 2001.

List of court cases

Note: Page numbers are given at the end of each court case.

Perry v. *Sidney Phillips & Son* (a firm) [1982] 1 *All ER* 1005, [1982] 3 *All ER* 705, [1982] 1 *WLR* 1297   191

Phillips v. *Price* [1959] Ch. 181   211

Pirelli General Cable Works Ltd v. *Oscar Faber & Partners* [1983] 1 *All ER* 65, [1983] 2 *WLR* 6   188

Polivitti Ltd v. *Commercial Union Assurance Co. Plc* (1987) 254 Re. J. 1990   230

Rich Investments Ltd v. *Cumgate Litho Ltd* [1988] *EGCS* 132   210

Rona v. *Pearce* [1953] 162 *EG* 380   190

Scammell v. *Dicker* (2005) *EWCA* Civ 405 and (2005) 42 *EG* 236   20

Sidnell v. *Wilson* 1966   210

Stevenson v. *Nationwide Building Society* [1984] 25 July 1984   197

Straudley Investments Ltd v. *Barpress Ltd* [1987] 282 *EG* 1224   207

Treml v. *Ernest W. Gibson & Partners* [1984] 27 June 1984   196

Watts and Another v. *Morrow* [1991] 14 *EG* 111   198

Whitehouse v. *Jordan* 1981   230

Yianni v. *Edwin Evans & Sons* (a firm) [1981] 3 *All ER* 592 [1981] 3 *WLR* 843   189

*Cases discussed in Chapter 19.

Useful addresses

Asbestos

Asbestos Removal Contractors Association
237 Branston Road, Burton upon Trent, Staffs, DE14 3BT, UK
Tel: 01283 531126
Web: www.arca.org.uk

Brickwork

Brick Development Association Limited
Woodside House, Winkfield, Windsor, Berkshire, SL4 2DX, UK
Tel: 01344 885 651
Web: www.brick.org.uk

Building Research

Building Research Establishment
Garston, Watford, Herts, WD25 9XX, UK
Tel: 01923 664000
Web: www.bre.co.uk

Builders

Federation of Master Builders
Gordon Fisher House, 14–15 Great James Street, London, WC1N 3DP, UK
Tel: 020 7242 7583
Web: www.fmb.org.uk

Cavity Insulation

National Cavity Insulation Association
PO Box 12, Haselmere, Surrey, GU27 3AH, UK
Tel: 01428 264011
Web: www.ncia-ltd.org.uk

Cavity Foam Bureau
PO Box 79, Oldbury, Warley, West Midlands, B69 4PW, UK
Tel: 0121 544 4949

Cement and Concrete

British Cement Association
Riverside House, 4 Meadows Business Park, Station Approach, Blackwater,
Camberley, Surrey, GU17 9AB, UK
Tel: 01276 608700
Web: www.cementindustry.co.uk

Damp Proofing

British Wood Preserving and Damp Proofing Association
1 Gleneagles House, Vernongate, Derby, DE1 1UP, UK
Tel: 01332 225100
Web: www.bwpda.co.uk

Decorators

Painting and Decorating Association
32 Coton Road, Nuneaton, Warwickshire, CV11 5TW, UK
Tel: 024 7635 3776
Web: www.paintingdecoratingassociation.co.uk

Electrical

National Inspection Council for Electrical Installation Contracting
Vintage House, 37 Albert Embankment, London, SE1 7UJ, UK
Tel: 020 7564 2323
Web: www.niceic.org.uk

Electrical Contractors Association
ESCA House, 34 Palace Road, London, W2 4HY, UK
Tel: 020 7313 4800
Web: www.eca.co.uk

Gas

CORGI
1 Elmwood, Chineham Park, Crockford Lane, Basingstoke, RG24 8WG, UK
Tel: 0870 401 2200
Web: www.corgi-gas-safety.com

Geological Maps

British Geological Society
Natural History Museum, Earth Galleries, Exhibition Road, London, SW7 2DE, UK
Tel: 020 7589 4090
Web: www.bgs.ac.uk

Glazing

Glass and Glazing Federation
44/48 Borough High Street, London, SE1 1XB, UK
Tel: 0870 042 4255
Web: www.ggf.co.uk

Heating and Ventilation

Heating and Ventilating Contractors' Association
Esca House, 34 Palace Court, London, W2 4JG, UK
Tel: 020 7313 4900
Web: www.hvca.org.uk

New Homes

National House-Building Council
Buildmark House, Chiltern Avenue, Amersham, HP6 5AP, UK
Tel: 01494 723530
Web: www.nhbc.co.uk

Zurich Municipal Building Guarantee Department
Southwood Crescent, Farnborough, Hampshire, GU14 0NJ, UK
Tel: 0870 2418050
Web: www.zurich.co.uk/Municipal/

Plastering

Federation of Plastering and Drywall Contractors
The Building Centre, 26 Store Street, London, WC1E 7BT, UK
Tel: 020 7580 3545
Web: www.fpdc.org.uk

Plumbing

Institute of Plumbing and Heating Engineering
64 Station Lane, Hornchurch, Essex, RM12 6NB, UK
Tel: 01708 472791
Web: www.iphe.org.uk

Roofing

National Federation of Roofing Contractors
24 Weymouth Street, London, W1G 7LX, UK
Tel: 020 7436 0387
Web: www.nfrc.co.uk

Structural Engineers

Institution of Structural Engineers
11 Upper Belgrave Street, London, SW1X 8BH, UK
Tel: 020 7235 4535
Web: www.istructe.org.uk

Surveyors

Royal Institution of Chartered Surveyors
Surveyor Court, Westwood Way, Coventry, CV4 8JE, UK
Tel: 0870 333 1600
Web: www.rics.org

Thatching

National Society of Master Thatchers Limited
13 Parkers Hill, Tetsworth, Thame, Oxfordshire, OX9 7AQ, UK
Tel: 01844 281208
Web: www.nsmt.co.uk

Thatching Advisory Services Ltd
The Old Stables, Redenham Park Farm, Redenham, Andover, Hampshire, SP11 9AQ, UK
Tel: 01264 773820
Web: www.thatchingadvisoryservices.co.uk

Timber

Timber Research and Development Association
Stocking Lane, Hughenden Valley, High Wycombe, HP14 4ND, UK
Tel: 01494 569600
Web: www.trada.co.uk

Trees

The Arboricultural Association
Ampfield House, Romsey, Hampshire, SO51 9PA, UK
Tel: 01794 368717
Web: www.trees.org.uk

Index